Attainment's EXPLORE math

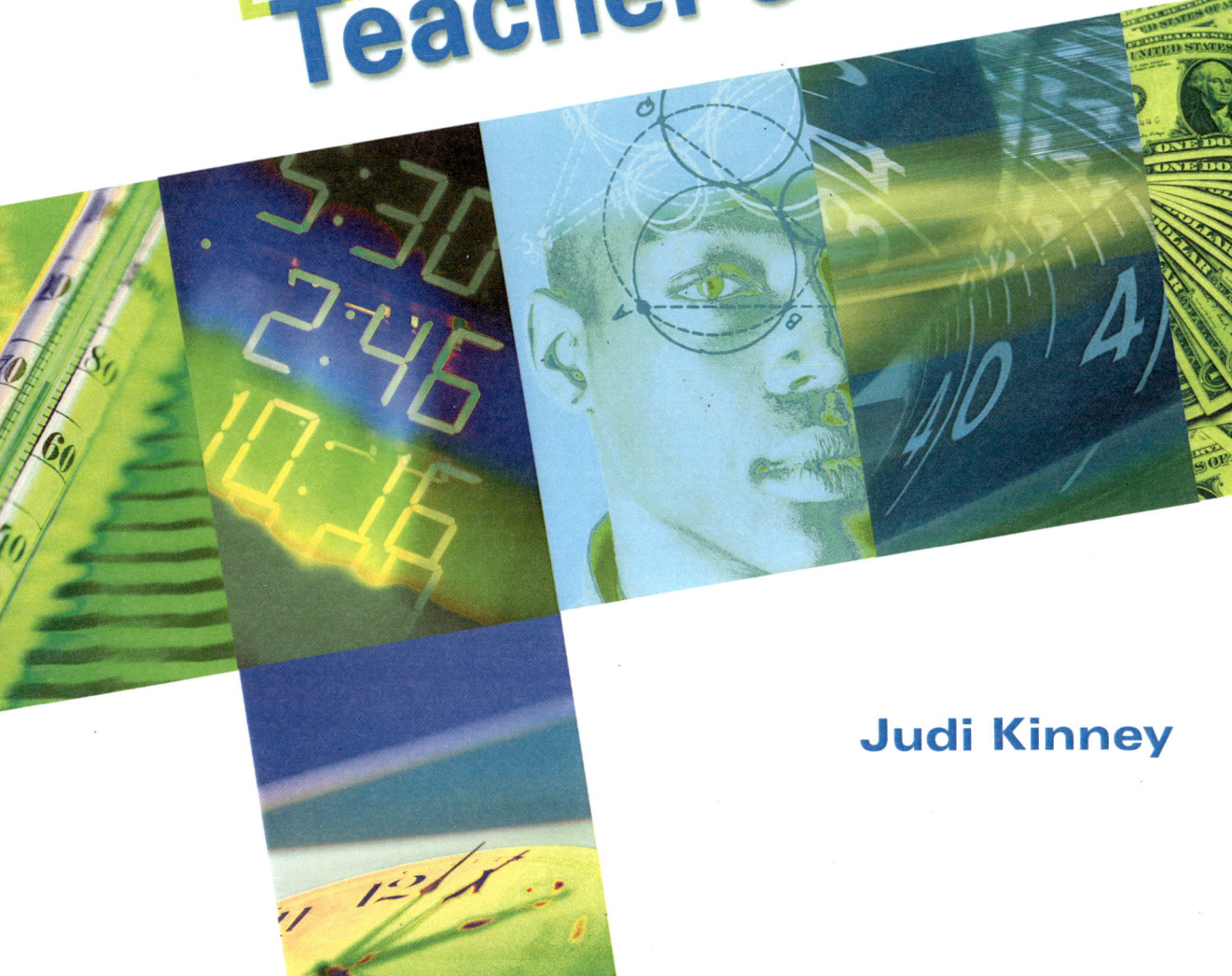

Teacher's Manual

Judi Kinney

Explore Math Win/Mac CD

This CD contains printable PDFs of the
Teacher's Guide and Student Workbook.
You can review and print pages from your computer.
The PDF (portable document format) file requires
Acrobat Reader software.

If you have Acrobat Reader already on your computer,
select and open the file from the CD.

To install Acrobat Reader:
Windows: Run ARINSTALL.EXE on the CD.
Mac: Run Reader Installer on the CD.

After installation, run Acrobat Reader and open
the file **EM_Manual.pdf** or **EM_Student.pdf** from the CD.

Explore Math Teacher's Manual

Judi Kinney, Author
Jo Reynolds, Graphic Design

An Attainment Company Publication
©2010 Attainment Company, Inc. All rights reserved.
Printed in the United States of America
ISBN 1-57861-696-4

Attainment Company, Inc.

P.O. Box 930160 • Verona, Wisconsin 53593-0160 USA
Phone: 800-327-4269 • Fax: 800.942.3865
www.AttainmentCompany.com

Reproducible resources within this material may be photocopied
for personal and educational use.

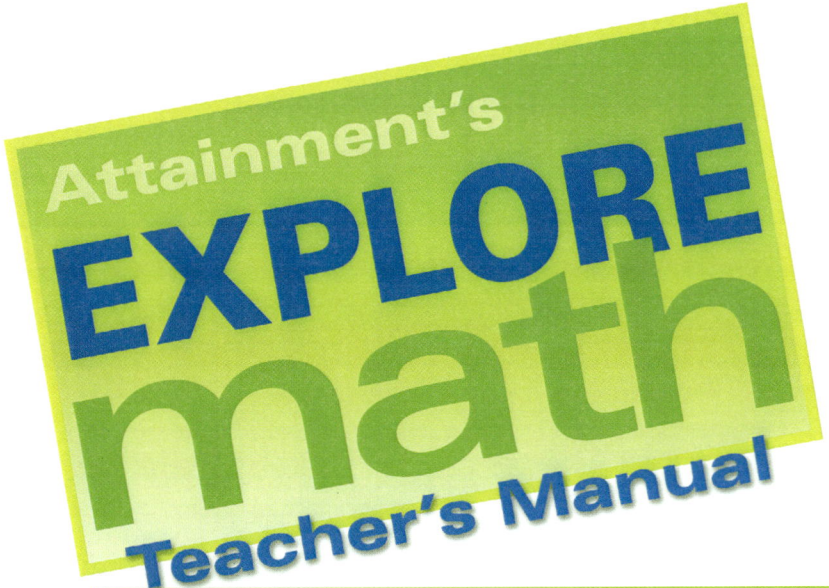

Table of Contents

Introduction 5

NTCM Standards 8

1 Vocabulary 9

2 0-12 15

3 0-18 43

4 0-100 71

5 0-1000 99

6 Fractions 119

7 Answer Key 143

8 Appendix 161

9 Bibliography 167

Introduction

Attainment's Explore Math books are designed to give students as many visual cues as possible to solve word problems. The books include a Teacher's Manual with suggested lessons for each workbook page and a Student Workbook. Many of the word problems include addition and subtraction computations, fractions, as well as problems that simulate everyday or real-life situations. There are six chapters in the books, starting with vocabulary words and ending with fractions. The chapters are organized so that the teacher can select workbook pages that are specific to individual student needs. Four of the chapters, 0–12, 0–18, 0–100, and 0–1000 are arranged using a similar format. If a teacher has students at different academic levels, she can assign pages that cover a concept but are at different levels of difficulty. For example, using the time telling pages in each chapter, one student may need to solve time to the hour in Chapter 0–12 and another may need to solve time problems using elapsed time to the hour.

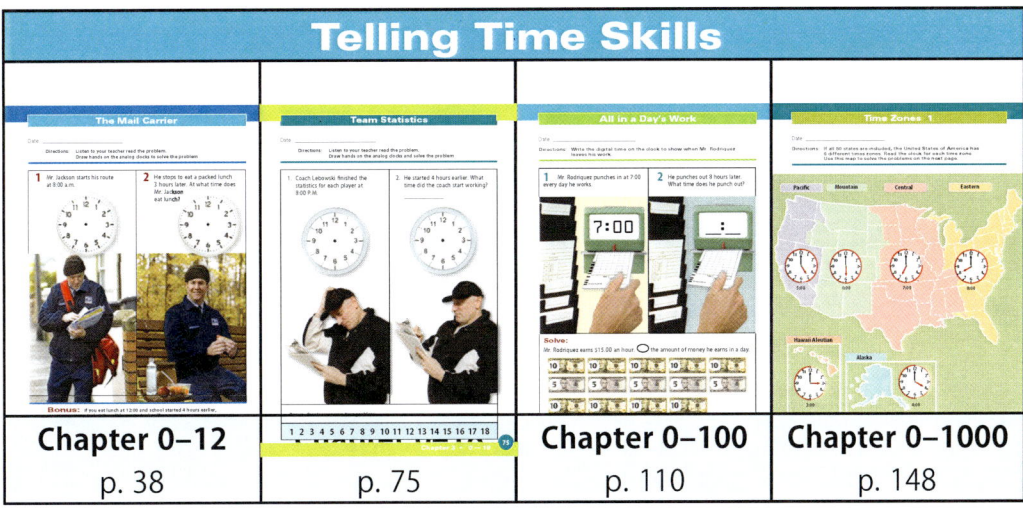

The following is a chart to help the teacher find pages in the student workbook that follow similar skills in order to help address the various academic abilities of students found in a special education classroom.

Skill	Chapter 0–12	Chapter 0–18	Chapter 0–100	Chapter 0–1000
matching	p. 18	p. 54	p. 91	pp. 128–129
maps	pp. 19–20	p. 55	pp. 92	pp. 130–131
addition	pp. 21–28	pp. 56–61	pp. 94–99	pp. 132–137
missing addends	pp. 29–30	pp. 62, 63, 66	p. 100–101	pp. 138–139
subtraction	pp. 31–35	pp. 67–74	pp. 102–109	pp. 140–143
time (clock)	pp. 36–38	pp. 75–78	pp. 110–113	pp. 144–146 pp. 148–149
time (calendar)	pp. 39–43	X	X	p. 147
money	pp. 44–45	pp. 79–81	pp. 114–117	pp. 150–151
graphs	pp. 46–49	pp. 82–85	pp. 118–123	pp. 152–153
math riddles	pp. 50–51	pp. 86–87	pp. 124–125	pp. 154–155

The Teacher's Manual

The Teacher's Manual has lesson plans for all of the worksheets in the Student Workbook. Each lesson plan has a Materials list that tells what worksheets and other items are used for that lesson.

These lesson plans allow the teacher to teach the skill before assigning the worksheet. Most of the lessons take about fifteen to twenty minutes to teach. Teachers should preview the lessons before introducing the worksheet. **Some lessons incorporate more than one worksheet because they cover the same skill.** The teacher can assign all of the worksheets in the lesson or spread the lesson over several days.

The Teacher's Manual has an Answer Key to all of the worksheets.

The Student Workbook

There are six chapters in the Student Workbook. The directions will need to be read by the teacher; the pictures on the page are clues to help the student solve the word problems.

When reading the directions, point out important words that can help the student figure out how to solve the problem, e.g., "altogether" or "how many were left." The teacher does not need to start at the beginning of the book and move sequentially through it to the end.

The **Vocabulary Chapter** consists of vocabulary words and worksheets to practice using the words found in the directions on the worksheets. The exercises in this chapter are designed to give students practice using important math words. Students can cut out and use the vocabulary word cards (pp. 6–9) as a word bank. If you need extra vocabulary word cards, make them by printing out from the PDF CD or by photocopying cards in the Appendix, pp. 161–166.

Chapters 0–12, 0–18, 0–100, and 0–1000 all follow a similar format. The format starts with a matching exercise, then some map work followed by addition and subtraction word problems. There are measurement problems, telling time problems (which focus mostly on elapsed time), money problems, and graphs to plot and interpret. At the end of each chapter are math riddles that involve learning how to predict an outcome, or use clues to solve a math riddle.

Number lines are at the bottom of the pages for Chapters 0–12 and 0–18. These number lines are a tool the students can use to solve the problems found in these two chapters. The **Hundreds Chart** at the beginning of Chapter 0–100 serves the same purpose. Students will need a **calculator** to solve or check the problems found in Chapter 0–1000.

The **Fractions Chapter** covers fractions of whole objects or fractions of sets. Again the teacher should select the pages which are appropriate for the students in the class.

NTCM Standards

Attainment's Explore Math Workbook is aligned to many NCTM Standards.

Number and Operations

Students use their knowledge of numbers and operations to solve everyday or real-life problems. They learn how to use whole numbers and fractions. They develop an understanding that fractions are parts of a whole and parts of a set.

Algebra

Students use pictorial representations to solve conventional problems. Students learn to analyze patterns and complete them.

Measurement

Measurement activities teach students important, everyday real-life skills. Students learn to solve problems that include, length, width, capacity, weight, temperature, money, and time.

Data Analysis

Students learn to plot and to interpret information found on graphs. Students learn to ask questions or make predictions based upon the data presented.

Problem Solving

Students learn to apply math computational skills to real-life situations and word problem formats. They learn to use a variety of strategies to solve the word problems.

Communications

Students learn to use mathematical language and apply it to pictures, graphs and math computations. They are encouraged to discuss, read, and write to express mathematical ideas found in word problems.

Connections

Students discover that mathematical ideas are connected and can be applied to real-life situations. Students are encouraged to use these concepts when relating findings to each other.

Chapter 1
Vocabulary

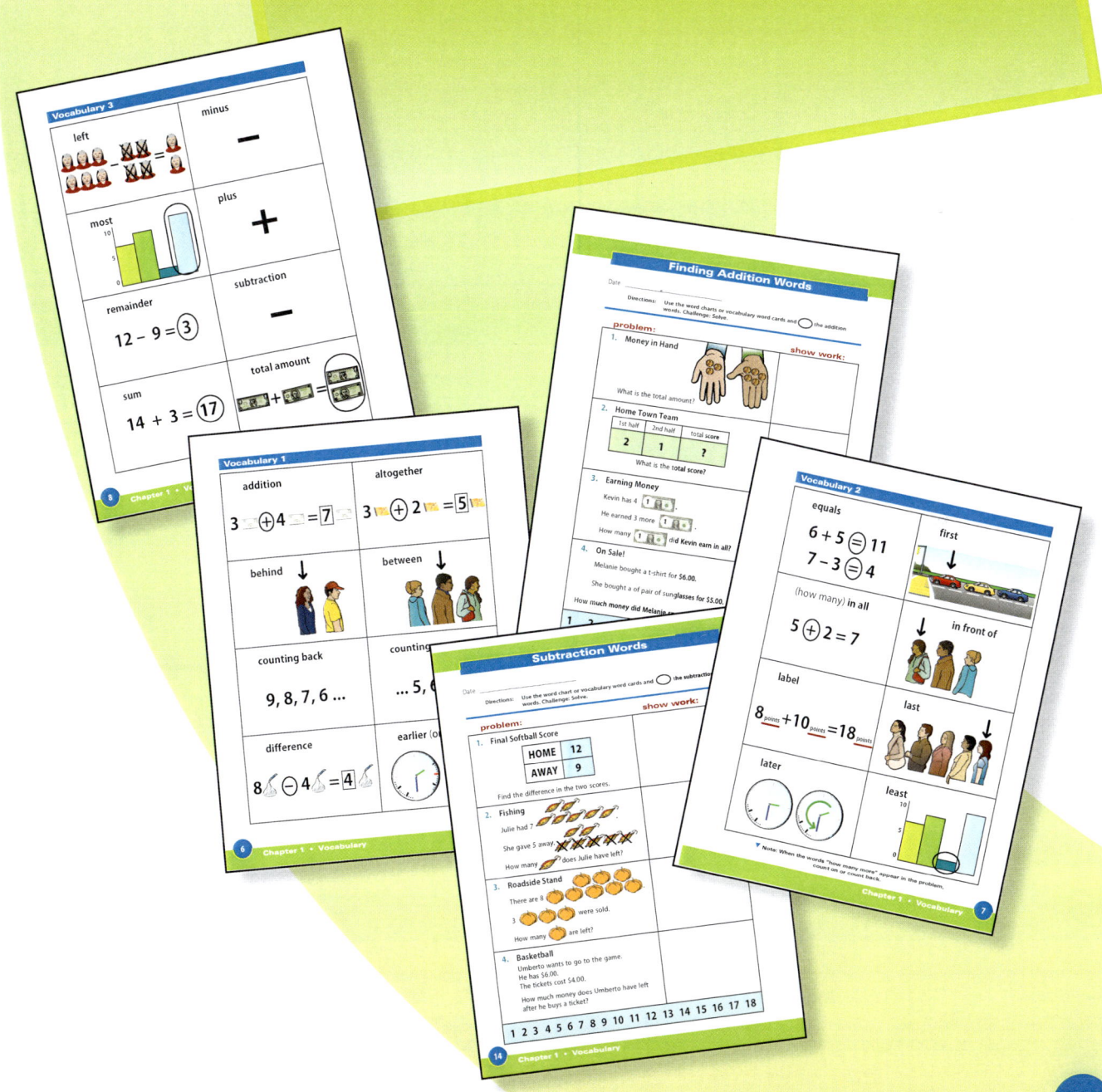

Chapter 1 • Vocabulary

Lesson 1

Objective

S. will read and define a set of vocabulary words.

Materials

- selected Word Card(s) (referenced in student workbook, pp. 6–9)
- dry board
- markers
- manipulatives

Procedure

1. Select a set of math vocabulary words for the students to learn.

2. Present one word at a time. Write the new word on the dry board.

3. Pointing under the word, read the word.

4. Next, point to the word and tell the students to read it with the teacher.

5. Demonstrate the meaning of the word using manipulatives.

6. Call on student volunteers to read the word independently and to define the word using the manipulatives.

7. Write the new word in a list of previously learned words, using the new word more than once.

8. Tell student volunteers to read the list of words.

9. Give a set of learned vocabulary Word Cards to each student to read and define.

10. If a student misses a word more than once, use the word as a new word to introduce in the next lesson.

11. When proficient, tell the students to use one of the blank cards and to define the word using their own words or illustrations.

Assign **selected vocabulary Word Cards.**

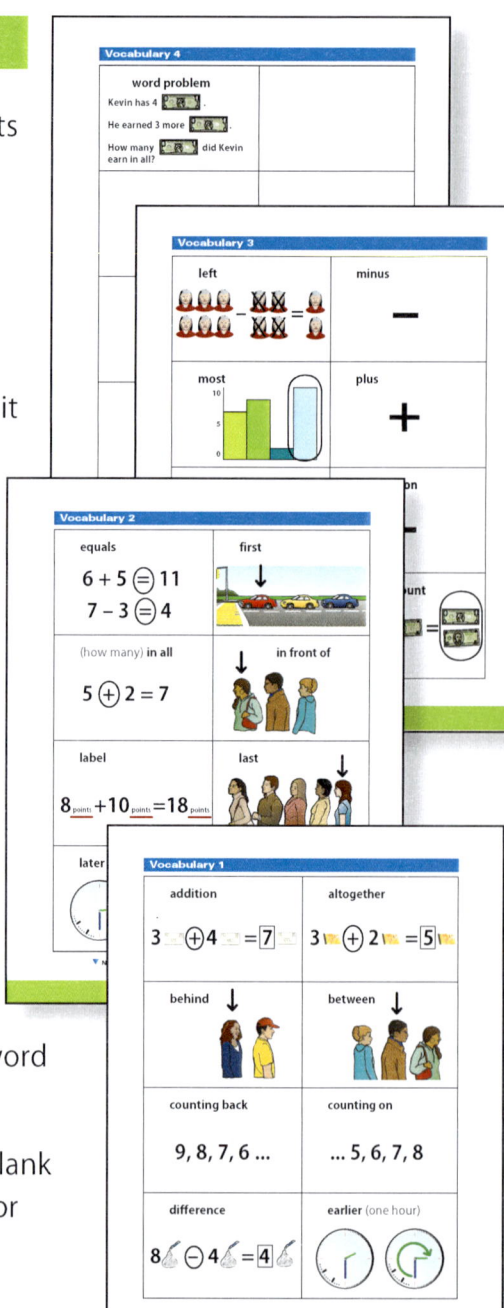

Lesson 2

Objective

Objective: S. will read a word(s) and state whether to add or subtract.

Materials

- selected Word Cards
- dry board
- markers
- teacher-made word problems on the board, with important math words underlined
- student word charts, pp. 10–11: Addition Words and Subtraction Words

Procedure

1. Point to and read the words on the Addition Words chart.
2. Lead students through the list again.
3. Tell students that each time they see one of the word(s) in a problem, they need to add.
4. Point to a problem. Read it. Point to the underlined word. Ask students what this word tells them to do.
5. Continue with other addition word problems.
6. When firm, introduce the Subtraction Words.
7. Follow the same procedure.
8. Tell students to use the charts to help them solve the word problems found in their workbook.
9. Blank lines are for words that either students or the teacher may want to add.

Note: Post **Addition Words** and **Subtraction Words** (photocopy or print out) to allow for easy student reference.

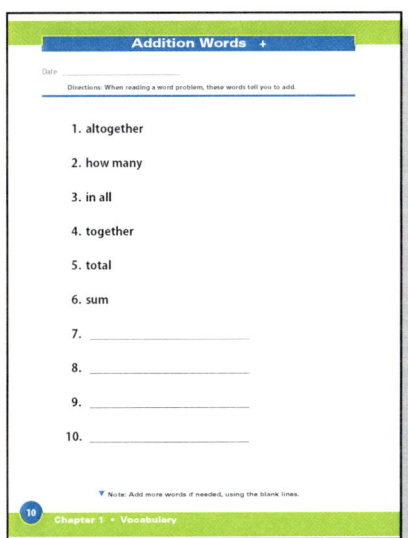

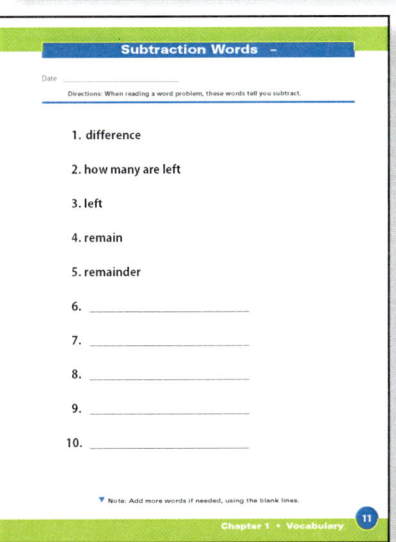

Chapter 1 ● Vocabulary

Lesson 3

Objective

S. will identify important position words in a math sentence.

Materials

- small manipulatives
- selected Word Cards
- pencils
- student worksheet, p. 12: Where Am I?

Procedure

1. Place a set of manipulatives in a row in front of the students.
2. Hold up a vocabulary card, e.g., **first.**
3. Read the word and point to the manipulative that is first in the row.
4. Lead students through the task.
5. Call on individual students to point to the first manipulative.
6. Continue with each additional word and until students are firm.

Assign the worksheet: **Where Am I?**

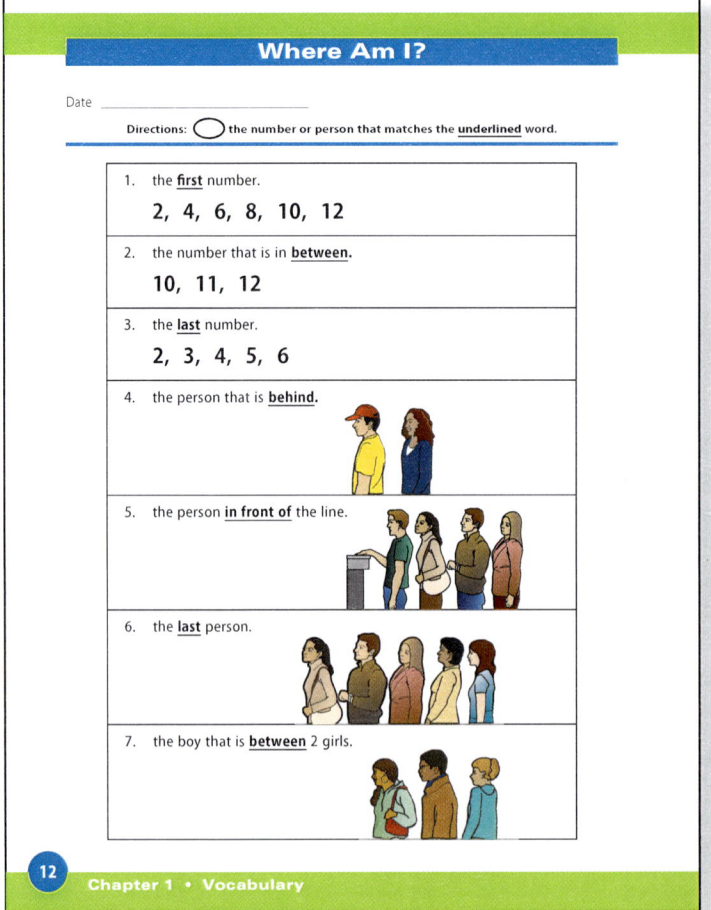

Lesson 4

Objective

S. will solve word problems using key words in the problem.

Materials

- selected Word Cards
- teacher-made word problems
- dry board
- markers
- posted addition and subtraction words
- pencils
- student worksheets, pp. 13–16: Finding Addition Words, Subtraction Words, Add or Subtract? 1 and 2

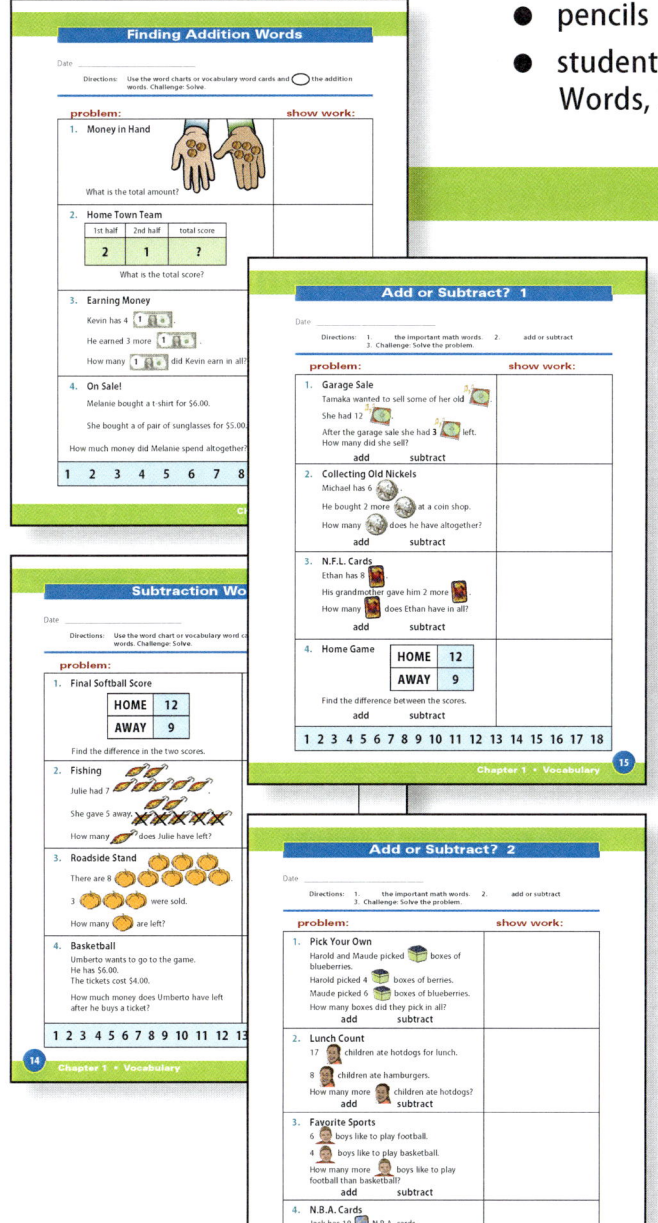

Procedure

1. Using the vocabulary cards, review reading the words.

2. Write a teacher-made word problem on the board using some of the learned words.

3. Underline the important addition and subtraction words.

4. Read the word problem.

5. Ask individual students to read the underlined word and tell whether to add or subtract.

6. Students can use word cards or the Addition and Subtraction word charts posted in the room or in their book to tell whether to add or subtract.

7. Continue until students are firm.

Assign worksheets:
Finding Addition Words, Subtraction Words, Add or Subtract? 1 and 2

Chapter 1 • Vocabulary

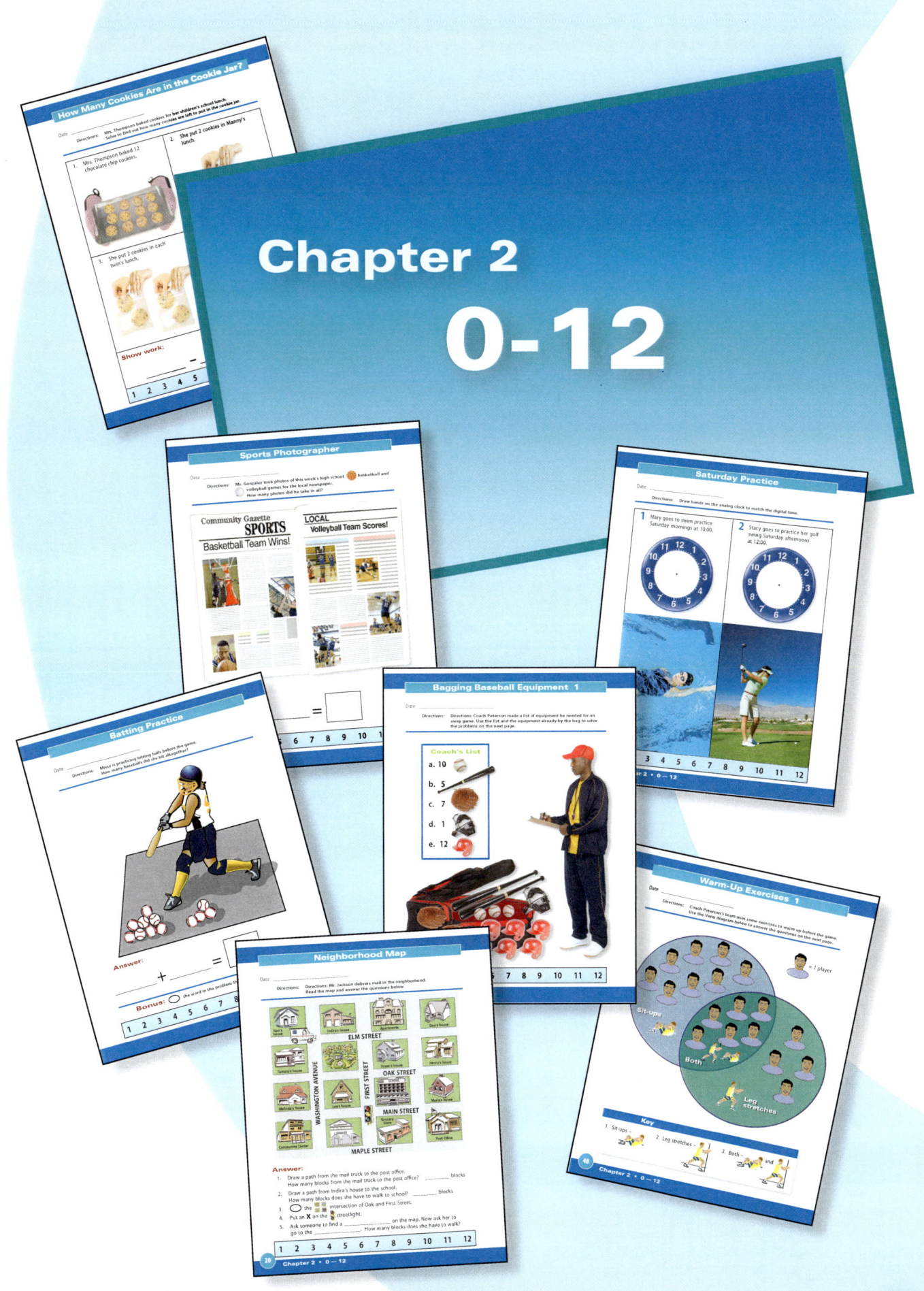

Lesson 1

Objective

S. will match the number of bills to the total price of an item.

Materials

- sets of dollar bills to 12 dollars
- a variety of items with price tags on each
- pencils
- student worksheet, p. 18: A Graphic Novel

Procedure

1. Students are to imagine going to a garage sale.

2. Assign student pairs, one to be a customer and the other the seller.

3. The seller lays various items in front of the customer to "buy."

4. Give the customer a set of dollar bills up to twelve dollars to use to buy items.

5. The customer counts the money.

6. He selects an item to buy.

7. He decides if he has enough to buy the item selected.

8. If he has enough money, the customer counts the money that matches the price of the item and gives it to the seller. If the customer doesn't have enough, he selects another item.

9. The seller counts the money to be sure that it matches the price of the item selected.

10. Students reverse roles.

11. Continue until the students are firm.

Assign the worksheet:
A Graphic Novel

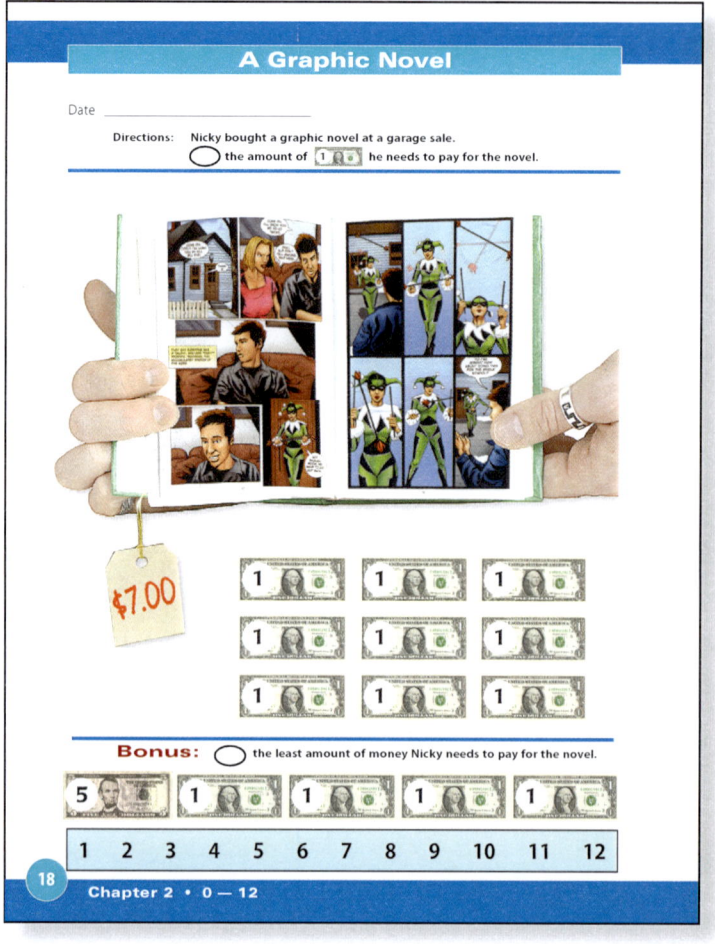

Explore Math Teacher's Manual

Lesson 2

Objective

S. will locate objects on a map.

Materials

- a teacher-made map of a street or hallway in a building, etc., with homes/rooms numbered by two, drawn on a dry board
- markers
- pencils
- student worksheet, p. 19: Elm Street

Procedure

1. Point to the map and tell students to brainstorm why maps are important.

2. Make a list of students' ideas.

3. Point out special features on the teacher-made map.

4. Ask individual students to find specific information using the map.

5. Continue until all the important features have been identified.

6. Write down the numbers 2–12 counting by two.

7. Students practice counting and writing numbers by two until proficient.

8. Erase addresses or room numbers in random order and tell student volunteers to write the correct address or room number on the map using their knowledge of counting by two.

Assign worksheet: **Elm Street**

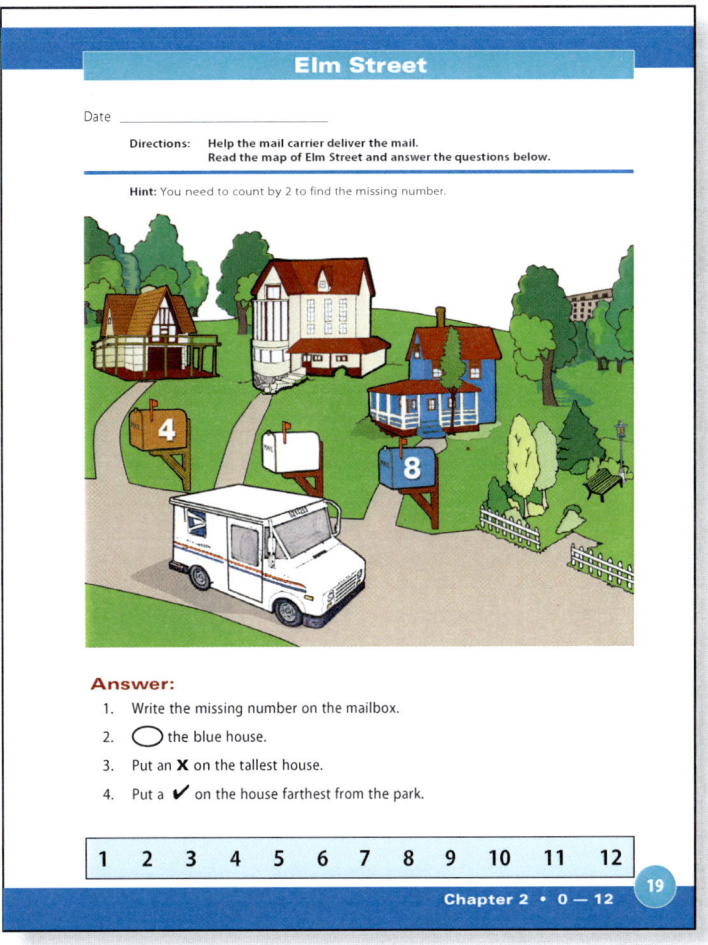

Lesson 3

Objective

Using a map, S. will count how many blocks there are from one location to another.

Materials

- a teacher-made map of a neighborhood on a dry board
- colored markers
- pencils
- student worksheet, p. 20: Neighborhood Map

Procedure

1. Review list of reasons why maps are important.
2. Point out special features of the teacher-made neighborhood map.
3. Ask individual students to find certain places on the map.
4. Demonstrate how to draw a path from one place to another.
5. Count how many blocks are walked to go from one place to another.
6. Create scenarios where a student must draw a path from one place to another on the map.
7. The student counts the blocks that are walked.
8. Continue until the students are firm.

Assign the worksheet:
Neighborhood Map

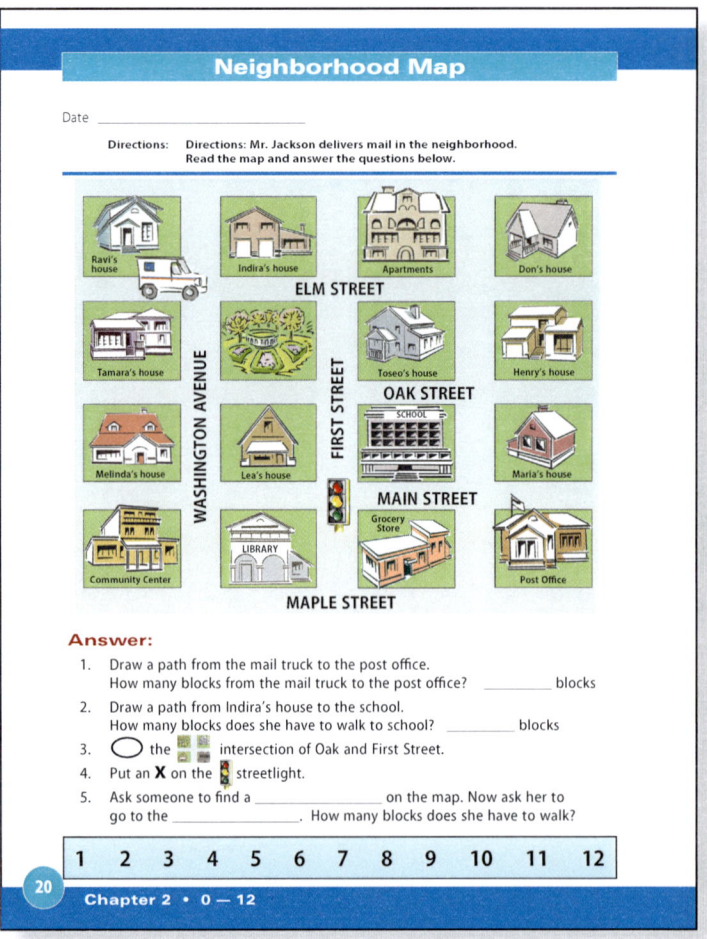

Lesson 4

Objective

S. will solve an addition word problem to twelve.

Materials

- teacher-made word problems
- dry board
- colored markers
- two small boxes
- a set of paper clips, coins, or other small manipulatives
- pencils
- student worksheet, p. 21: The Mail Carrier

Procedure

1. Write **altogether** on the board.

2. Write examples of word problems that students must solve to find out how many items there are altogether.

3. Tell students to circle the word altogether in the problems.

4. Lead students to discover that **altogether** means to add.

5. Place two boxes in front of the students.

6. Create several scenarios where students must count the manipulatives in each box and write an addition problem to solve how many items there are altogether, e.g., a bank clerk puts 6 coins in one box and 4 coins in another. How many coins are there altogether?

7. When firm, encourage students to create at least one problem using the same materials.

Assign the worksheet: **The Mail Carrier**

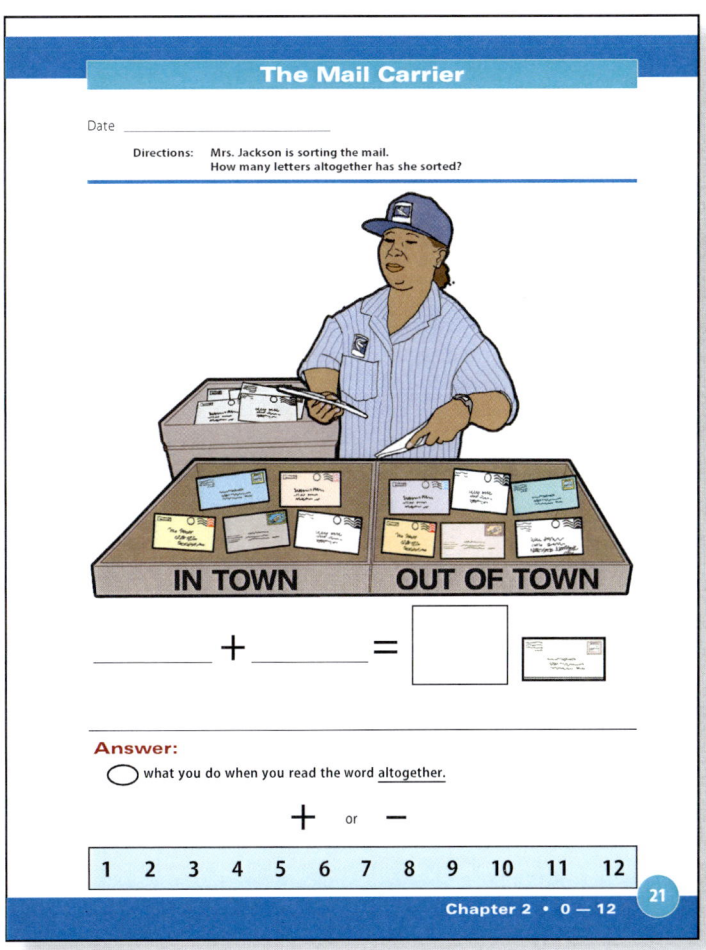

Lesson 5

Objective

S. will solve an addition word problem to twelve.

Materials

- a set of paper clips, coins, or other small manipulatives
- pencils
- student worksheet, p. 22: Loading the Mail Truck

Procedure

1. Review with students the meaning of the word altogether when found in a math word problem.

2. Create several scenarios where students must count two sets of objects and solve to find out how many there are altogether, e.g., The teacher corrected 6 math papers and 3 spelling papers. How many papers were corrected altogether?

3. When firm, encourage students to create at least one problem, using the same materials.

Assign the worksheet:
Loading the Mail Truck

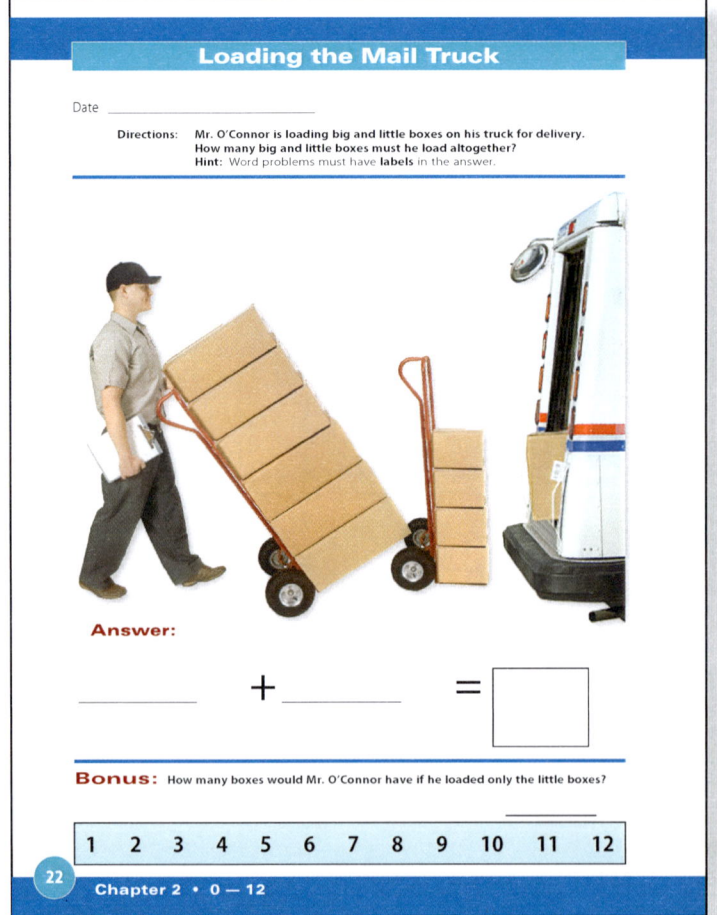

Lesson 6

Objective

S. will solve an addition word problem to twelve.

Materials

- teacher-made word problems on the board with the word altogether in them
- a set of paper clips, coins, or other small manipulatives
- pencils
- student worksheet, p. 23: Batting Practice

Procedure

1. Write several word problems on the board.

2. Read the problems one at a time and ask student volunteers to find the word altogether in the problem. The student circles the word and states that the word altogether means to add.

3. Create several scenarios where students must count two sets of objects, and solve to find out how many there are altogether, e.g., Sara gave 6 sugar cookies to Derrick and 6 chocolate chop cookies to Mary. How many cookies altogether did she give away?

4. When firm, encourage students to create at least one problem using the same materials.

Assign the worksheet: **Batting Practice**

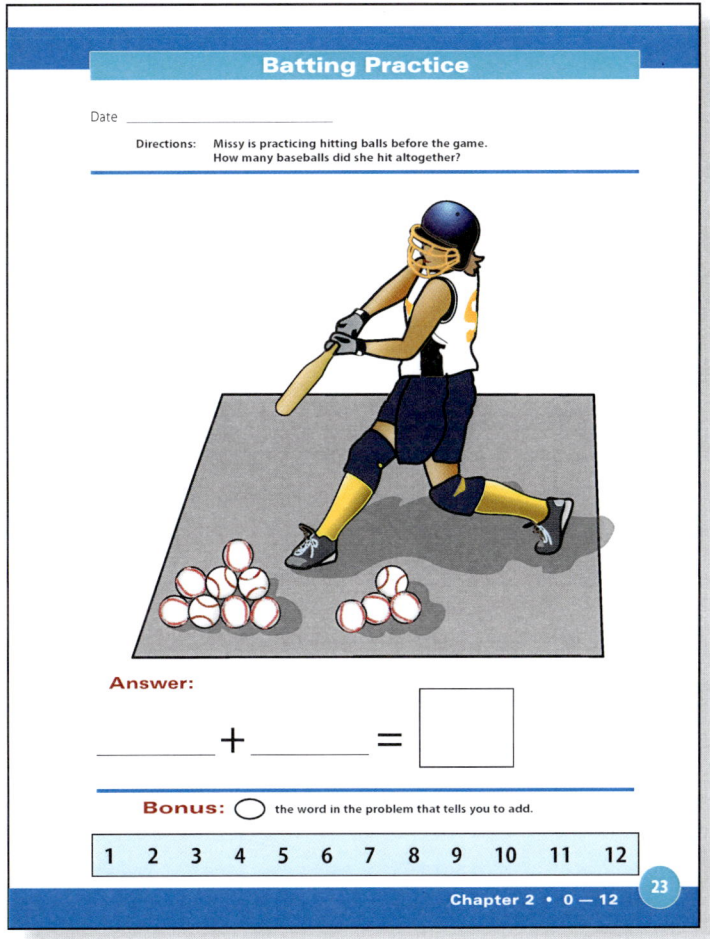

Chapter 2 • 0—12

Lesson 7

Objective

S. will solve an addition word problem to twelve.

Materials

- teacher-made word problems
- sports pages from a newspaper
- dry board
- colored markers
- pencils
- student worksheet, p. 24: Sports Photographer

Procedure

1. Write the words **in all** on the board.

2. Write examples of word problems that students must solve to find out how many items there are **in all**.

3. Lead students to discover that **in all** means to add.

4. Using the newspaper photos, create several addition story problems that students must solve using the photos. Students must use addition words **in all** in the problem.

5. Encourage the students to use labels in their solutions to the problems.

6. Continue until students are firm.

Assign the worksheet:
Sports Photographer

Lesson 8

Objective

S. will solve an addition word problem to twelve.

Materials

- teacher-made scoreboard on a white board or chalkboard
- markers/chalk
- pencils
- student worksheet, p. 25: Who Won?

Procedure

1. Talk about how scores can be made in a school football game.

2. Create different scenarios for scores.

3. Demonstrate how to solve to find the final score by adding the scores for the first and second half of a game.

4. Provide several examples where student volunteers have to add to find the final score for each team.

5. Students must tell what team won the game.

6. Emphasize that word problems need labels, and that a short way or abbreviation for writing point is **pt.**

7. Continue until students are firm.

Note: Students could play a game such as bouncing a ball. Each student gets two chances to bounce the ball. Record the bounces. The student must stop when he misses a bounce. Add the first and second chances to find the total amount a student could bounce a ball. Compare results.

Assign the worksheet: **Who Won?**

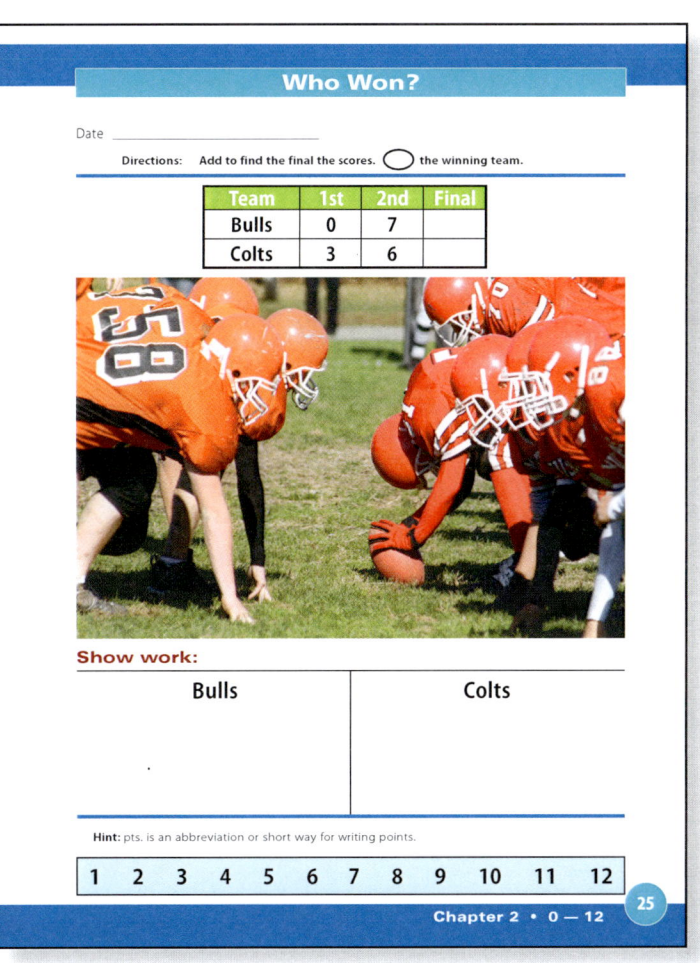

Lesson 9

Objective

S. will solve a word problem to twelve.

Materials

- teacher-made list or inventory of objects
- sets of manipulatives which match the teacher list
- a paper bag
- dry board
- markers
- pencils
- student worksheets, pp. 26-27: Bagging Baseball Equipment 1 and 2

Procedure

1. Make a list on the board.

2. Place some items in the paper bag that match the teacher-made list.

3. Create a scenario where the items in the bag must total the number on the list, e.g., Point to the list and say, "The art teacher needs 10 paintbrushes." Count the paintbrushes in the bag. "She has 7 brushes; how many more brushes does she need to match the number on the list?" Count up from seven: "8, 9, 10." Say: "I counted three more brushes. Seven plus three equals ten. The art teacher needs 3 more brushes." Place the 3 brushes in the bag.

4. Lead the students through the problem.

5. Call on individual students to solve the problem.

6. Continue until all of the objects on the list have been used.

Assign the worksheets:
Bagging Baseball Equipment 1 and 2

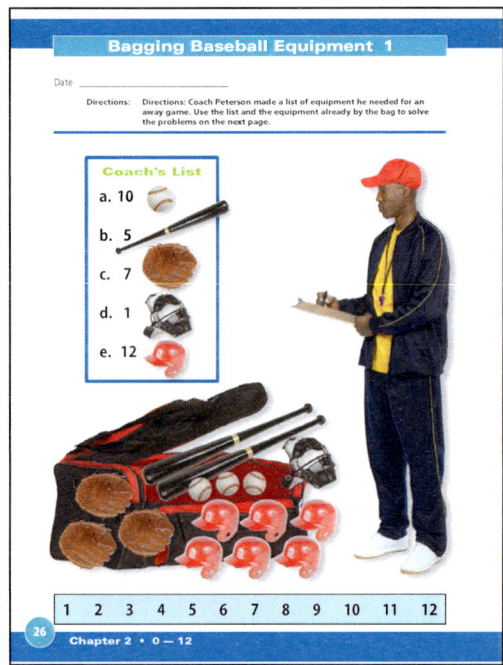

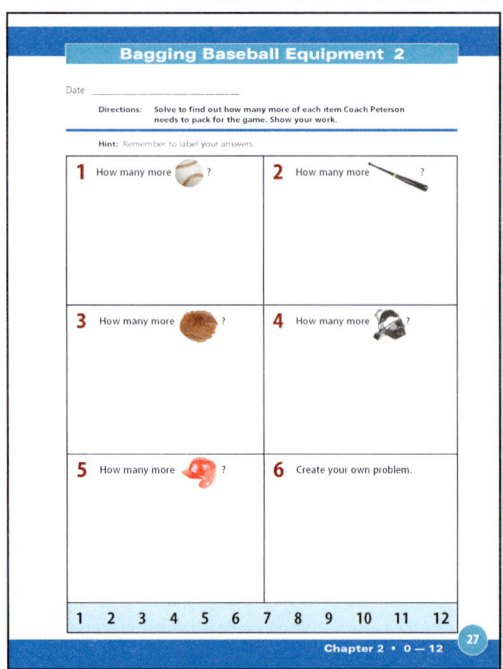

Explore Math Teacher's Manual

Lesson 10

Objective

S. will solve a word problem using column addition.

Materials

- teacher-made beanbag toss, magnetic dartboard, or other game-like board
- dry board
- markers
- pencils
- student worksheet, p. 28: Throwing Darts

Procedure

1. Draw a beanbag or dartboard, or use a magnetic board.

2. Create several scenarios using the game board by shading the numbers until three numbers have been used. For example: Margaret threw a dart and it landed on the number 3, the next dart landed on 5, and the final dart landed on 3. What was her total score? Add the three numbers to find the sum total.

3. Lead the students through the problem.

4. Continue until the students are firm.

Assign the worksheet:
Throwing Darts

Lesson 11

Objective

S. will solve a word problem using column addition with a missing addend.

Materials

- teacher-made column addition problems with one missing addend
- dry board
- markers
- sets of dollar bills to 12 dollars
- pencils
- student worksheets, pp. 29-30: Shooting Baskets and Tip Money

Procedure

1. Write several column addition problems on the board with a missing addend in each problem.

2. Point to a box with the missing addend.

3. Create a scenario for the problem.

4. Demonstrate how to solve the problem.

5. Lead the students through the process of solving the problem.

6. Continue until students are firm.

Assign worksheets:
Shooting Baskets and **Tip Money**

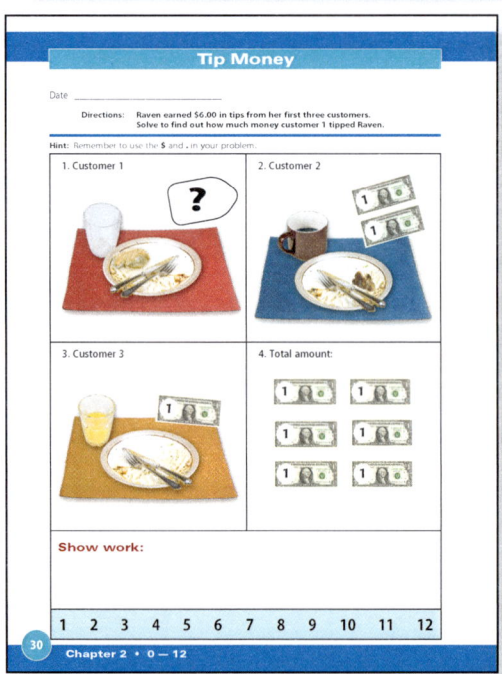

Lesson 12

Objective

S. will solve a subtraction word problem using money to twelve cents.

Materials

- teacher-made list of items and prices for sale
- dry board
- markers, set of 12
- pennies
- pencils
- student worksheet, p. 31: Day-Old Bake Sale

Procedure

1. Write a list of items and prices on the board.

2. Give a set of pennies to a student and ask her to count the pennies.

3. The student must "purchase" something from the list that does not exceed the amount of money in her hand.

4. The student purchases the item by giving the teacher the total amount of pennies needed to buy the item.

5. The student counts how much money she has left.

6. The student writes out the problem using cents as a label.

7. Repeat until the students are firm.

Assign worksheet:
Day Old Bake Sale

Lesson 13

Objective

S. will solve a subtraction word problem using money to twelve dollars.

Materials

- a set of items to buy, with price tags using the dollar and cents sign
- a set of 12 dollar bills
- pencil
- student worksheet, p. 32: A Sale on Sunglasses

Procedure

1. Lay the items to buy in a row in front of the students.

2. Give a student volunteer a set of dollar bills.

3. The student must select an item "to buy" that does not exceed the amount on the price tag.

4. The student purchases the item and counts out how much she has left in her hand.

5. The student writes out the problem using the dollar sign and decimal point.

6. Repeat with another student until everyone has had an opportunity to "purchase" an item.

Assign worksheet:
A Sale on Sunglasses

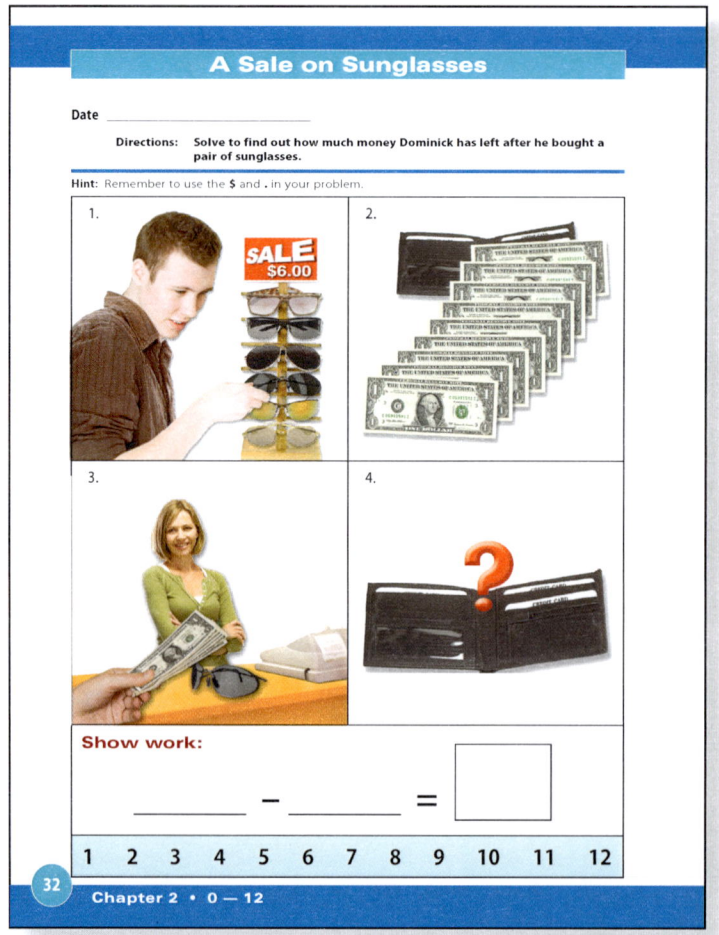

Lesson 14

Objective

S. will solve a word problem using addition or subtraction to twelve.

Materials

- sets of small manipulatives
- pencils
- student worksheet, p. 33: How Many Cookies Are in the Cookie Jar?

Procedure

1. Lay out a set of manipulatives in a row.

2. Create scenarios where the students must take away several sets of the manipulatives in order to find out how many are left.

3. Students can use counting up or counting back strategies, e.g., Terry had 10 candy bars and he gave 1 to Ed and 2 to Isaac. How many did he have left? Demonstrate how to solve the problem.

4. Lead the student through the problem as students say and write the problem.

5. Continue until students are firm.

Assign the worksheet:
How Many Cookies are in the Cookie Jar?

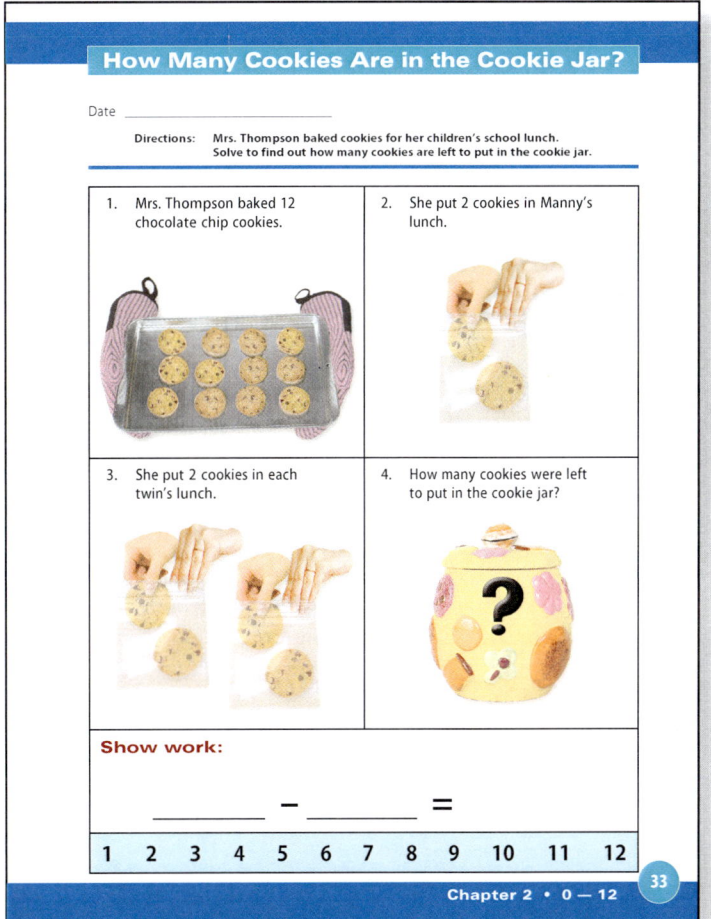

Lesson 15

Objective

S. will solve to find out how much money is left after making a purchase.

Materials

- a set of dollar bills to 12 dollars
- items "to purchase" with price tags
- pencils
- student worksheets, pp. 34-35: A Trip to the Mall 1 and 2

Procedure

1. Review how to read price tags with the dollar sign and decimal point.

2. Lay out items "to purchase".

3. Create a series of scenarios with different items to buy and amount of money to spend.

4. Demonstrate how to purchase an item.

   a. Select an item.

   b. Count the money in a wallet to be sure that there is enough to pay for the item.

   c. If yes, take the item to the checkout counter. If no, select another item.

   d. Count out the money to pay for the item.

   e. Count how much money is left.

5. Call on individual students to make a purchase and count the money that is left.

6. Continue until students are firm.

Assign the worksheets:
A Trip to the Mall 1 and **2**

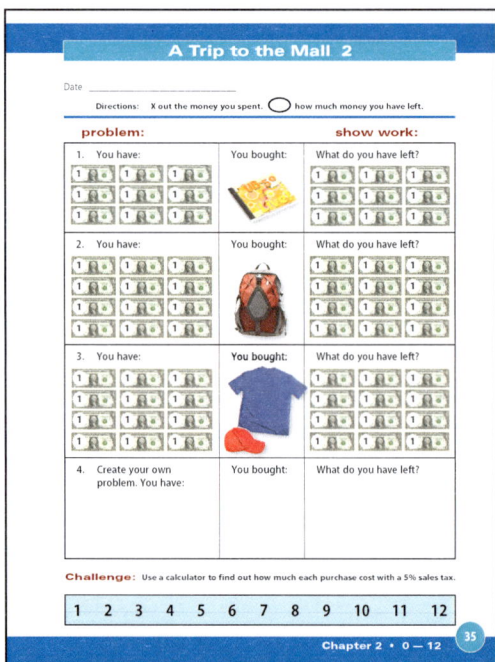

Lesson 16

Objective

S. will draw hands to the hour on an analog clock that matches the digital time.

Materials

- analog clock(s)
- teacher-made list of digital times to the hour
- dry board
- markers
- pencils
- student worksheet, p. 36: Saturday Practice

Procedure

1. Point to a digital time on the teacher-made list.

2. Review how to set an analog clock that matches the time.

3. Give individual students practice using the other digital times on the list.

4. Continue until the students are firm.

Assign the worksheet:
Saturday Practice

Note: Make a clock on the board like the one below to teach practice counting around by 5.

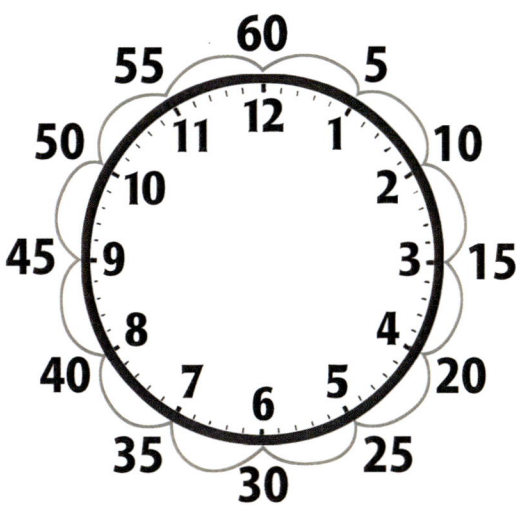

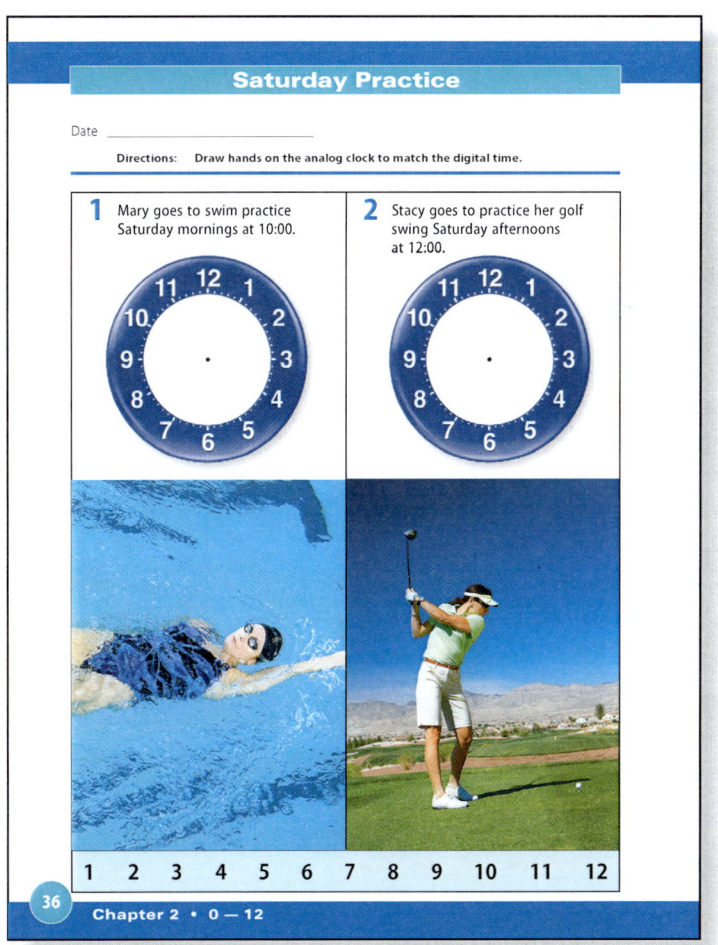

Lesson 17

Objective

S. will draw hands on an analog clock that is an hour later than a stated time.

Materials

- analog clock(s)
- teacher-made list of digital times to the hour
- dry board
- markers
- pencils
- student worksheet, p. 37: Guitar Practice

Procedure

1. Point to a digital time on the teacher-made list.

2. Set the analog clock to the time.

3. Create a time problem where something takes one hour to complete, e.g., a homework assignment, a long walk, etc.

4. Demonstrate how to set the analog clock an hour later.

5. Count around the clock by fives to one hour later.

6. Lead the students through the task.

7. Call on individual students to follow the same procedure.

8. Continue until the students are firm.

Assign the worksheet:
Guitar Practice

Explore Math Teacher's Manual

Lesson 18

Objective

S. will draw hands on an analog clock that is more than an hour later.

Materials

- analog clock(s)
- teacher-made lists of digital times
- dry board
- markers
- pencils
- student worksheet, p. 38: The Mail Carrier

Procedure

1. Review how to set a clock to an hour.

2. Review how to set the clock to one hour later.

3. Point to a digital time on the board.

4. Create a scenario so that the clock must be set at a time that is more than one hour later.

5. Lead students through the task.

6. Ask individual students to repeat the procedure.

7. Continue until the students are firm.

Assign the worksheet:
The Mail Carrier

Lesson 19

Objective

S. will read a daily calendar and answer questions based upon that day's activities

Materials

- examples of a daily calendar
- teacher-made calendar
- dry board
- markers
- pencils
- student worksheet, p. 39: A Daily Calendar

Procedure

1. Show examples of a daily calendar, both digital and paper if possible.

2. Brainstorm with students reasons for keeping a daily calendar.

3. Demonstrate how to fill out the calendar on the board.

4. Ask questions about the calendar and call on individual students to respond.

5. Continue until students are firm.

Optional: Tell students to keep a daily calendar for one day. Students can compare daily schedules and ask questions based upon individual calendars.

Assign the worksheet:
A Daily Calendar

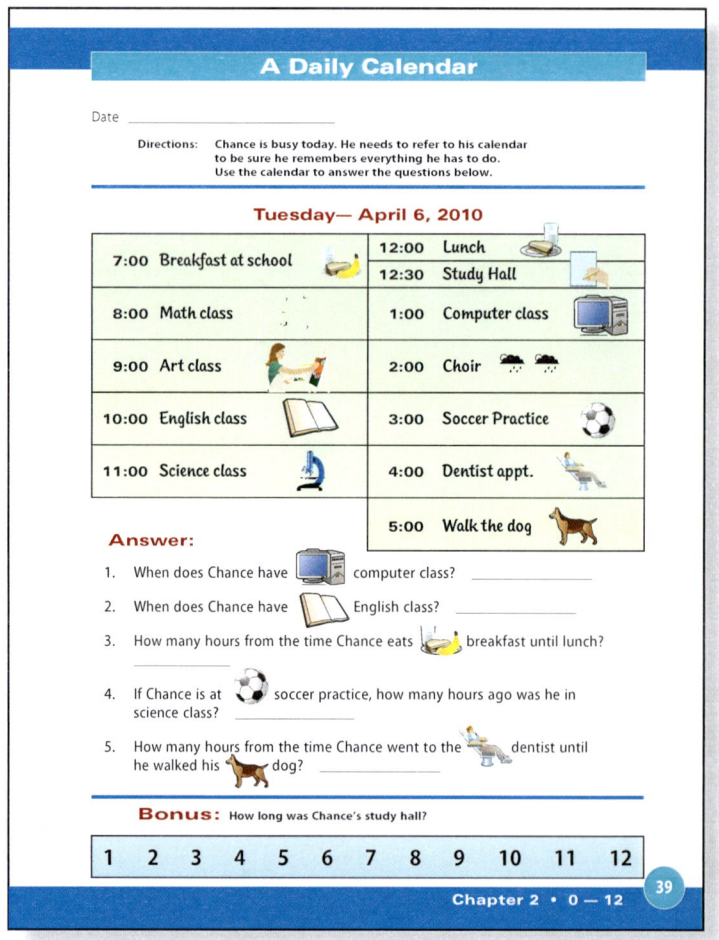

Lesson 20

Objective

S. will read a weekly calendar and answer questions based upon the week's activities.

Materials

- examples of a weekly calendar, both paper and digital
- teacher-made calendar
- dry board
- markers
- pencils
- student worksheets, pp. 40-41: Mrs. Ling's Calendar 1 and 2

Procedure

1. Show examples of a weekly calendar, both digital and paper.//
2. Brainstorm with students reasons for keeping a weekly calendar.
3. Demonstrate how to fill out a weekly calendar on the board.
4. Ask questions about the calendar and call on individual students to respond.
5. Continue until students are firm.

Assign the worksheets: **Mrs. Ling's Calendar 1** and **2**

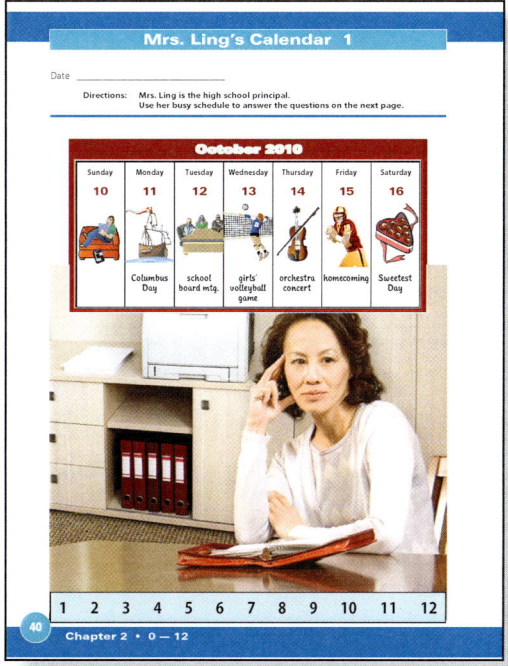
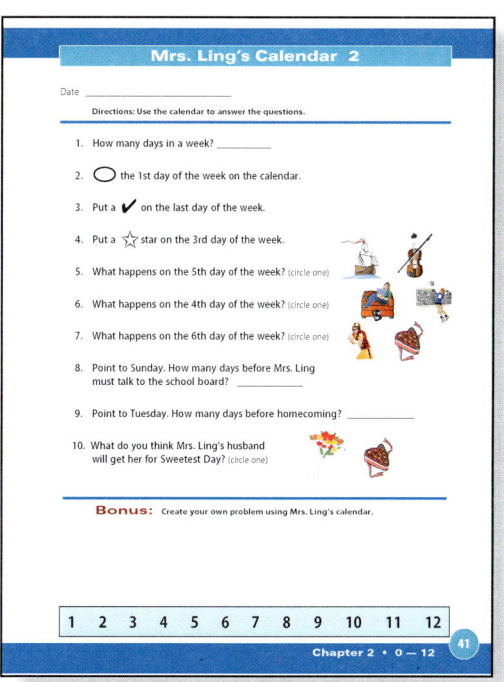

Lesson 21

Objective

S. will read a yearly calendar and answer questions based upon the year's activities.

Materials

- examples of a yearly calendar
- teacher-made calendar
- dry board
- markers
- pencils
- student worksheets, pp. 42-43: A Yearly Calendar 1 and 2

Procedure

1. Show examples of a yearly calendar, both digital and paper if possible.
2. Brainstorm reasons for keeping a yearly calendar.
3. Brainstorm what kind of jobs people have that require a yearly calendar.
4. Demonstrate how to fill out the teacher-made yearly calendar.
5. Ask questions about the calendar and call on individual students to respond.
6. Continue until students are firm.

Assign the worksheets: **A Yearly Calendar 1** and **2**

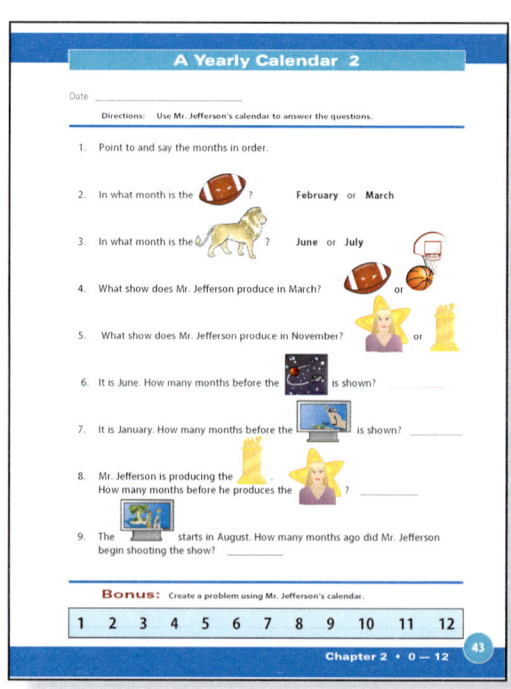

Explore Math Teacher's Manual

Lesson 22

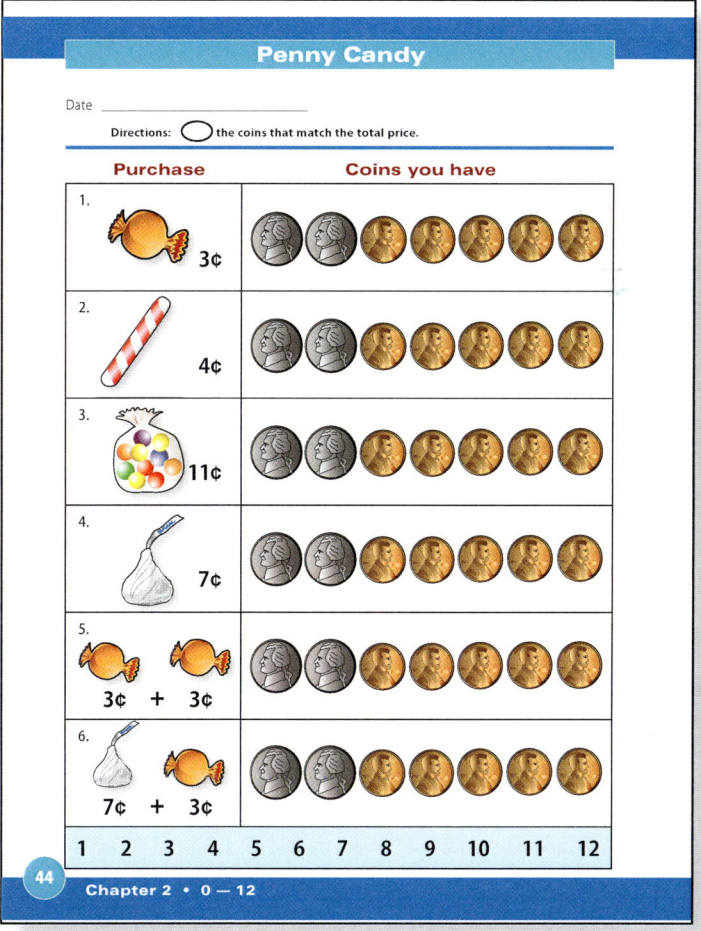

Objective

S. will match the purchase price of an item using nickels and pennies to twelve cents.

Materials

- small items to purchase with price tags to 12 cents
- sets of nickels and pennies to 12 cents
- pencils
- student worksheet, p. 44: Penny Candy

Procedure

1. Place the items in a row in front of the students to purchase.
2. Select an item and a set of coins.
3. Count out the coins to match the price of the item. Use the smallest number of nickels and pennies.
4. Repeat the procedure.
5. Give individual students a set of coins and ask each to purchase an item.
6. Prompt the student to use the fewest amount of coins.
7. Continue until all students have had an opportunity to purchase an item.

Assign the worksheet:
Penny Candy

Lesson 23

Objective

S. will count the change back from a purchase price.

Materials

- small items to purchase with price tags to 12 cents
- sets of nickels and pennies
- pencils
- student worksheet, p. 45: Penny Candy Again

Procedure

1. Place items to purchase in a row in front of the students.

2. Count out a set of money and use the money to purchase an item. The item cannot cost more than the amount of money in the set, but can cost less than the amount.

3. Give the "clerk" the money. The "clerk" counts back the change from the purchase.

4. Assign student pairs, one is the customer and the other is the sales clerk.

5. The customer gets a set of coins and uses them to purchase an item.

6. The customer gives the clerk the coins and the clerk counts back change.

7. Students switch roles and continue until they can count back change without prompts.

Assign the worksheet:
Penny Candy Again

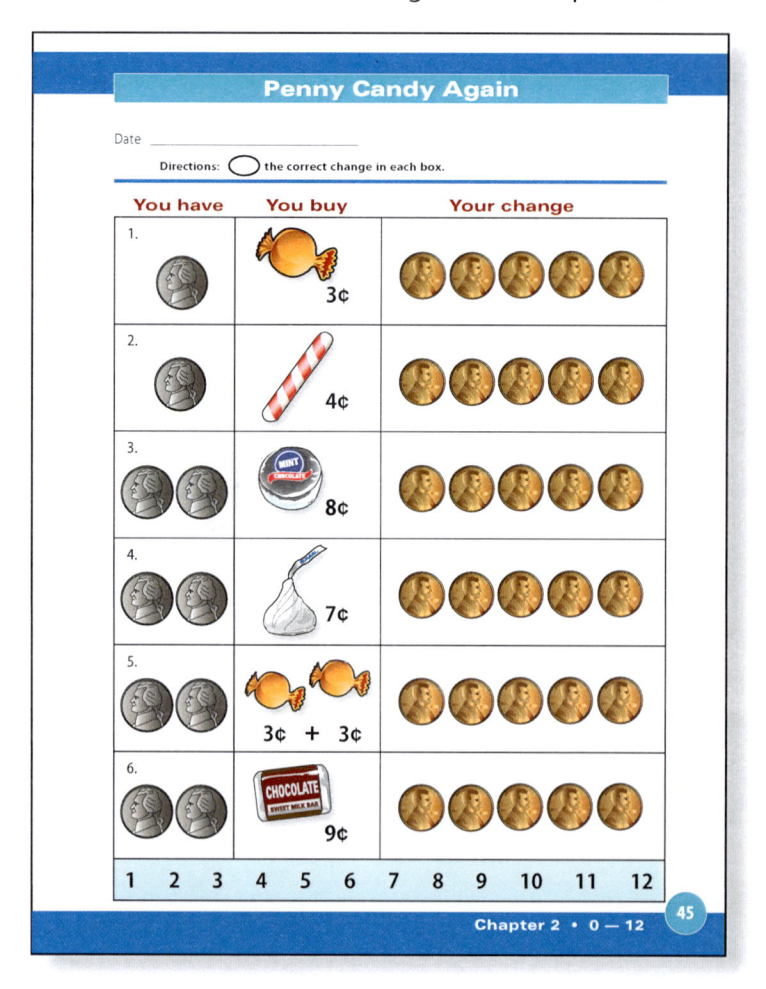

Lesson 24

Objective

S. will interpret information from a pictograph.

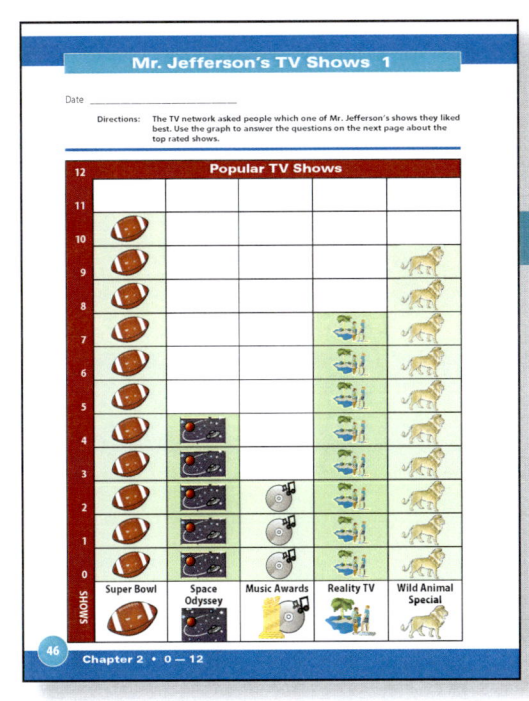

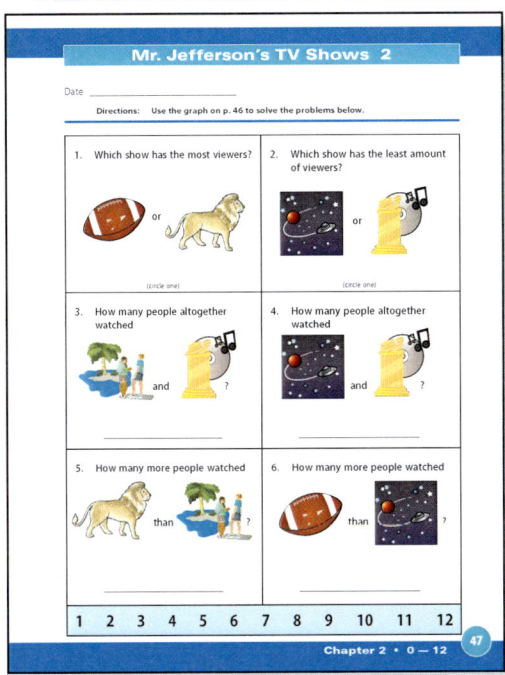

Materials

- examples of pictographs from newspapers, etc.
- a class-generated pictograph
- dry board
- markers
- pencils
- student worksheets, pp. 46-47: Mr. Jefferson's TV Shows 1 and 2

Procedure

1. Show examples of pictographs.

2. Tell students to brainstorm ideas to make a picture graph.

3. Select one.

4. Create the graph on the board, prompting students to name the correct parts, e.g., title, labels for x- and y-axes, etc.

5. After the graph has been created, ask the students questions that will enable them to interpret the information or data on the graph.

6. Continue until all columns of the graph have been used.

Assign the worksheets:
Mr. Jefferson's TV Shows 1 and 2

Lesson 25

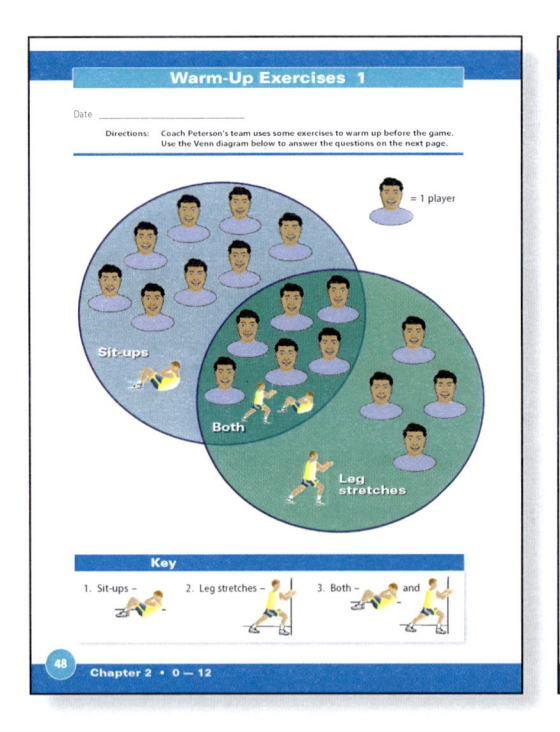

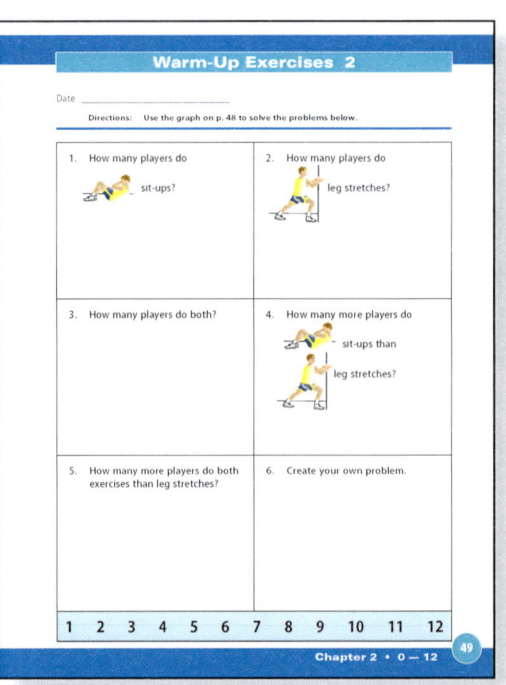

Objective

S. will interpret information from a Venn diagram using pictures.

Materials

- class-generated Venn diagram
- dry board
- markers
- pencils
- student worksheets, pp. 48-49: Warm-up Exercises 1 and 2

Procedure

1. Tell students to brainstorm ideas for a Venn diagram.

2. Select one.

3. Create the Venn diagram, prompting students where to label each part.

4. After the diagram has been created, ask the students questions about it.

5. Continue until all of the data has been used.

Assign the worksheets:
Warm-Up Exercises 1 and **2**

Lesson 26

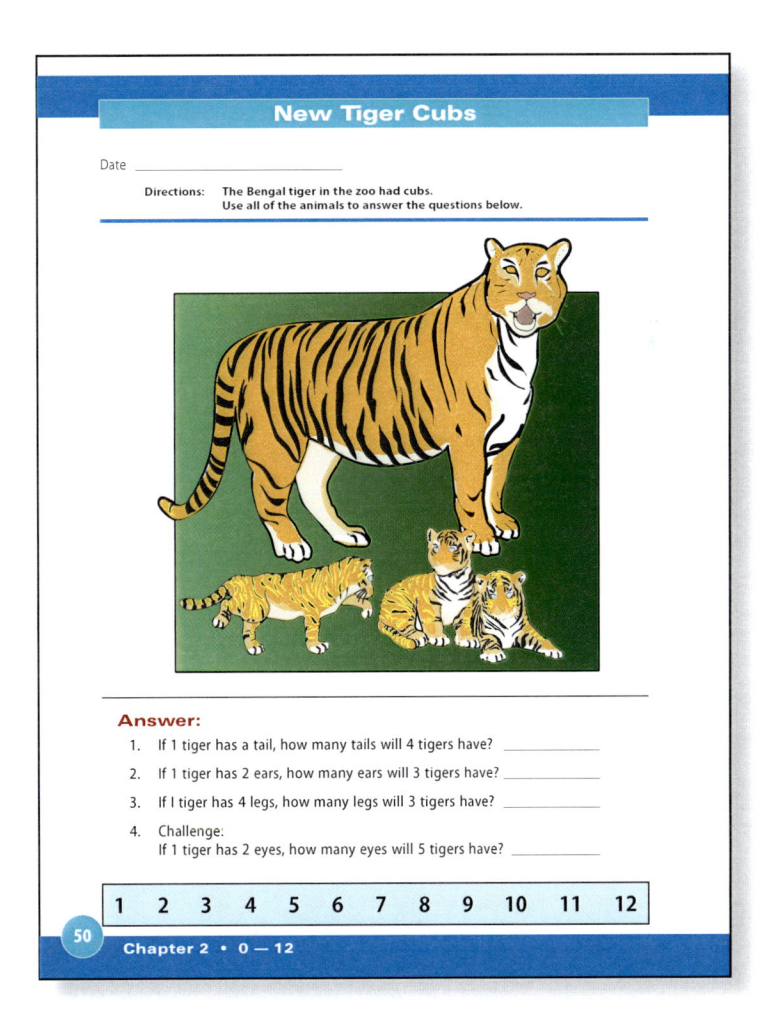

Objective

S. will skip count to solve a math riddle.

Materials

- assorted pictures of animals or people from magazines
- pencils
- student worksheet, p. 50: New Tiger Cubs

Procedure

1. Select a picture.

2. Using the picture ask students questions, e.g., If there are 5 people in the picture, how many eyes do they have altogether?

3. Continue questioning until the appropriate information has been asked.

4. Select another picture and continue procedure until students are firm.

Assign the worksheet:
New Tiger Cubs

New Tiger Cubs

Date _____

Directions: The Bengal tiger in the zoo had cubs. Use all of the animals to answer the questions below.

Answer:
1. If 1 tiger has a tail, how many tails will 4 tigers have? _____
2. If 1 tiger has 2 ears, how many ears will 3 tigers have? _____
3. If 1 tiger has 4 legs, how many legs will 3 tigers have? _____
4. Challenge:
   If 1 tiger has 2 eyes, how many eyes will 5 tigers have? _____

| 1 | 2 | 3 | 4 | 5 | 6 | 7 | 8 | 9 | 10 | 11 | 12 |

Lesson 27

Objective

S. will predict the results of a coin toss.

Materials

- a penny for each student
- student worksheet, p. 51: Coin Flip

Procedure

1. Tell students that they will learn to make and check a prediction.

2. Explain that a prediction is an educated, or good, guess about what may happen. A prediction is not always correct.

3. Give several examples of predictions.

4. Pass out coins and the worksheet: **Coin Flip**.

5. Tell students to predict whether the coin they flip will land more on heads or tails. Have them circle their choice.

6. Explain how to use the Tally Chart on the worksheet.

7. Students complete the worksheet and compare predictions.

Assign the worksheet:
Coin Flip

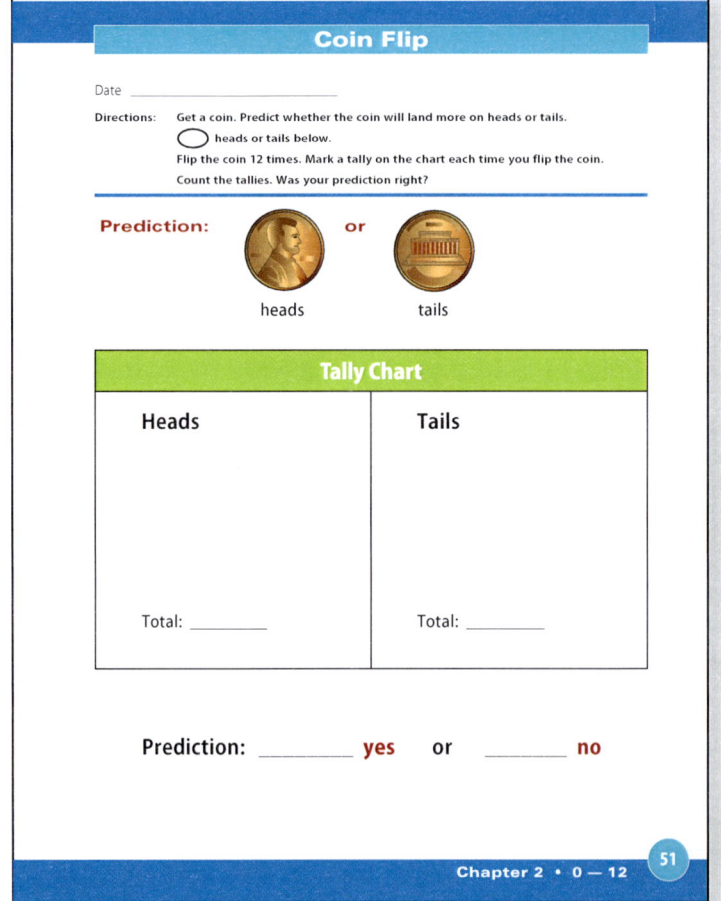

Chapter 3
0-18

Lesson 1

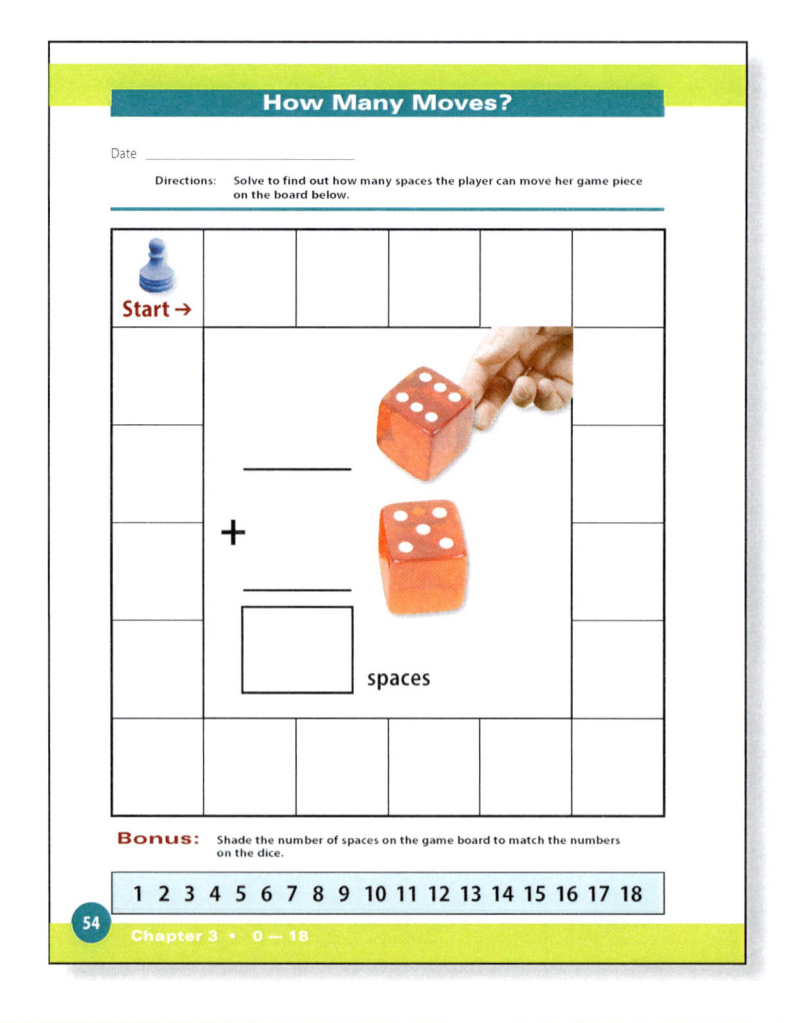

Objective

S. will match the number of dots on a pair of dice and add to find the total amount.

Materials

- a pair of dice
- game pieces
- game board
- pencils
- student worksheet, p. 54: How Many Moves?

Procedure

1. Roll a pair of dice and review with students how to count the total amount shown.

2. Lead the students through the task.

3. Roll the pair of dice again and move a game piece on the board that corresponds to the amount thrown.

4. Give individual students opportunities to do the same task.

5. Continue until students are firm.

Assign the worksheet:
How Many Moves?

Lesson 2

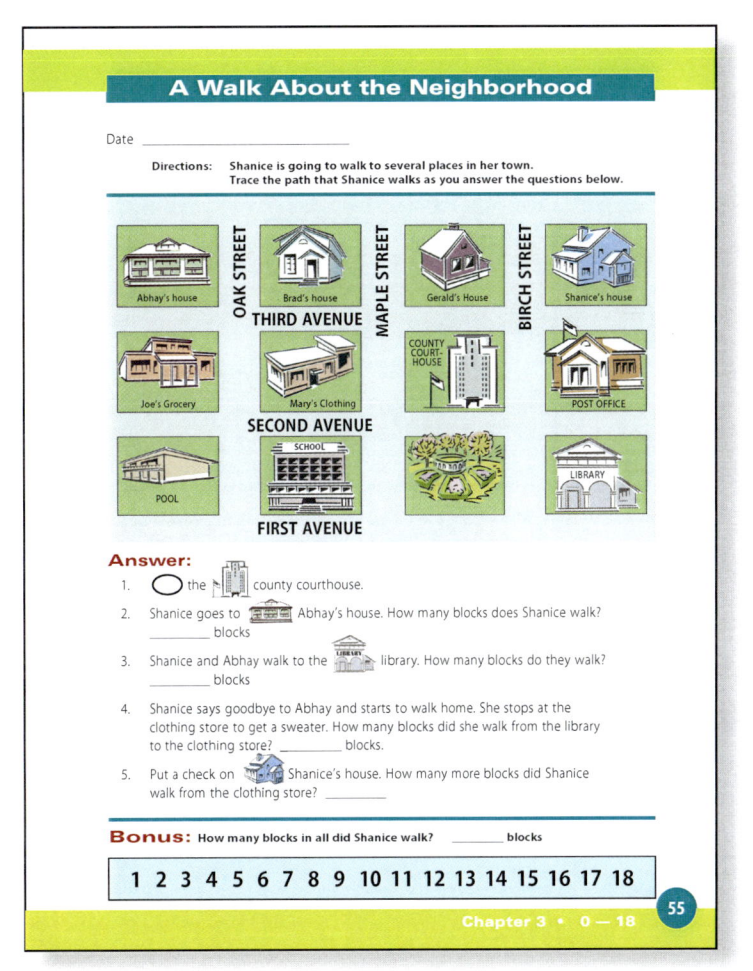

Objective

Using a map, S. will count how many blocks there are from one location to another.

Materials

- teacher-made map of a neighborhood on a dry board
- colored markers
- pencils
- student worksheet, p. 55: A Walk About the Neighborhood

Procedure

1. Point out special features on the map.

2. Ask students to identify places and streets on the map.

3. Give scenarios where students must move from one location to another using the map.

4. Students count the blocks as they move from one location to another. Encourage students to take the shortest route.

5. Continue until all students have had an opportunity to use the teacher-made map.

Assign the worksheet:
A Walk About the Neighborhood

Lesson 3

Objective

S. will solve an addition word problem to eighteen.

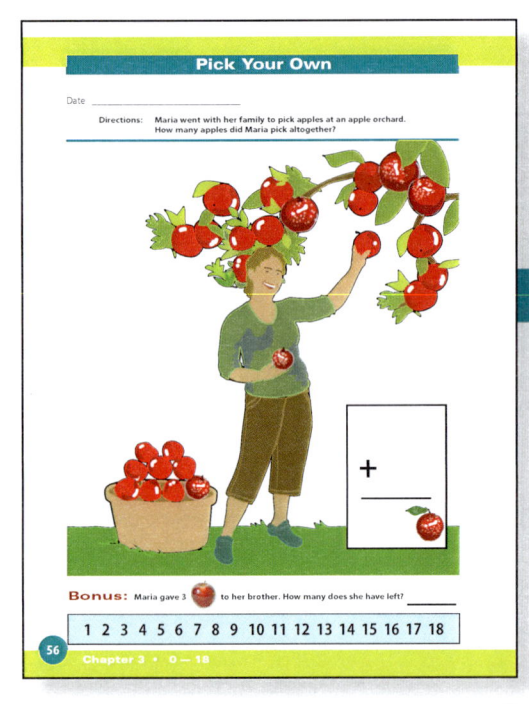

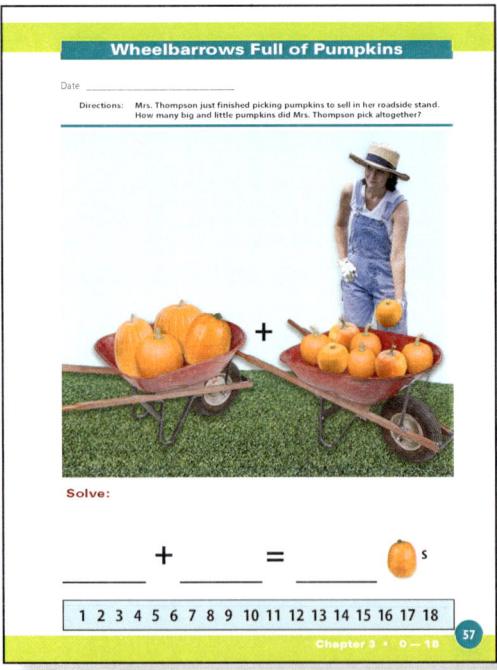

Materials

- teacher-made addition word problems using the word "altogether"
- dry board
- markers
- two small boxes
- a set of small manipulatives
- pencils
- student worksheets, pp. 56–57: Pick Your Own and Wheelbarrows Full of Pumpkins

Procedure

1. Read the first word problems written on the board.

2. Point to the word **altogether** and ask a student to read the word.

3. Prompt students to tell what operation they must perform when they read the word **altogether.**

4. Continue the procedure until all of the problems have been read.

5. Place two small boxes in front of the students.

6. Create several scenarios where students must count the manipulatives in each box and write an addition problem to solve how many items there are altogether.

7. Remember to reinforce the concept of labels when solving a word problem.

8. Continue until students are firm.

Assign the worksheets: **Pick Your Own** and **Wheelbarrows Full of Pumpkins**

Lesson 4

Objective

S. will solve an addition problem to eighteen.

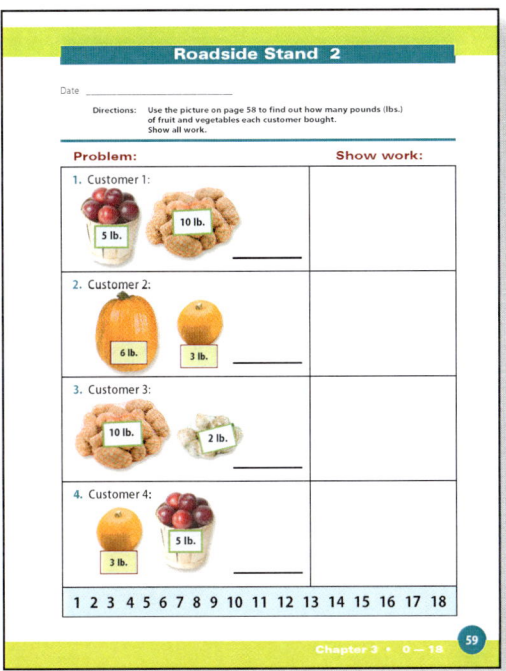

Materials

- small scale
- items to weigh (must be large enough to weigh in pounds)
- dry board
- markers
- pencils
- student worksheets, pp. 58–59: Roadside Stand 1 and 2

Procedure

1. Review with students how to weigh items on the scale.

2. As the items are weighed, review with students how to write weights using pounds.

3. Point out that **lb.** is a short way or abbreviation for writing pounds.

4. Select two items and weigh them one at a time.

5. Read the weight and write the weights in pounds on the board.

6. Add the two weights emphasizing that the label is in pounds.

7. Call on individual students to weigh two different items.

8. Continue until students can weigh the items and solve to find the total weight without prompts.

Assign the worksheets:
Roadside Stand 1 and **2**

47

Chapter 3 • 0–18

Lesson 5

Objective

S. will purchase two items and find the total number of dollars for the purchase.

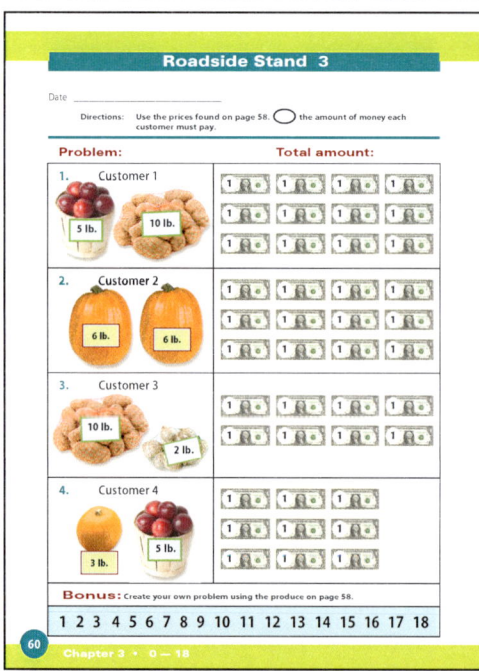

Materials

- sets of dollar bills to 18
- items used for weighing in Lesson 4, with price tags on them
- dry board
- markers
- pencils
- student worksheets, pp. 58 and 60: Roadside Stand 1 and 3

Procedure

1. Lay the items to buy in a row in front of the students.
2. Pick up a set of bills and count the money.
3. Select two items that total the amount.
4. The final amount cannot exceed the amount of bills in the set.
5. Count out the amount of bills that equal the total amount of the two items.
6. Pay for the items.
7. Clerk recounts the money to be sure that it is accurate.
8. Assign student pairs, one to be a customer and one the clerk.
9. The student pairs follow the same procedure.
10. When complete, students switch roles.
11. Continue until students are firm.

Assign the worksheets: **Roadside Stand 1** and **3**

Explore Math Teacher's Manual

Lesson 6

Objective

S. will solve for the total amount of a purchase using the dollar sign and decimal point.

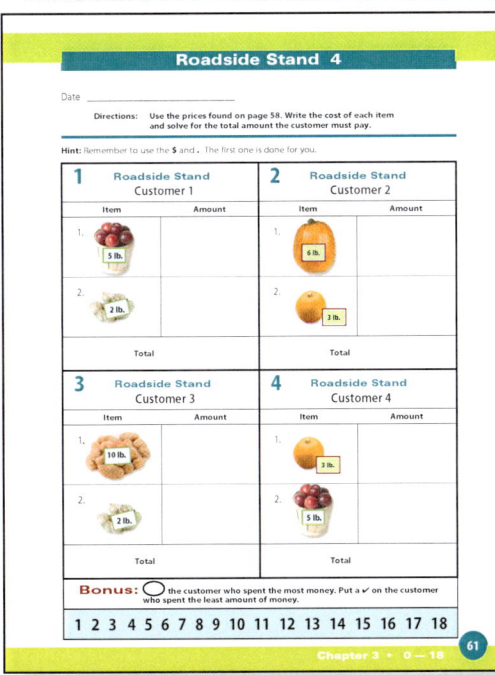

Materials

- sets of dollar bills to 18
- items used for weighing in Lesson 4, with price tags on them
- teacher-made sales receipts on the board
- dry board
- markers
- pencils
- student worksheets, pp. 58 and 61: Roadside Stand 1 and 4

Procedure

1. Explain to the students that another way to find out the total amount of a purchase is to write a sales receipt.

2. Select two items. Write the cost of one item on the first line of a receipt on the board.

3. Write the second item on the next line.

4. Add the total amount.

5. Count out the money from a set that equals the amount of the two items.

6. Assign a student pair to follow the same procedure.

7. Using the receipt on the board, one student is the customer and selects two items to buy.

8. The second student is the clerk, who writes the prices of the items and adds the total amount.

9. The customer pays for the items.

10. Students switch roles.

11. Repeat with another pair of students until all students have followed the procedure.

Assign the worksheets: **Roadside Stand 1** and **4**

Lesson 7

Objective

S. will solve for a missing addend.

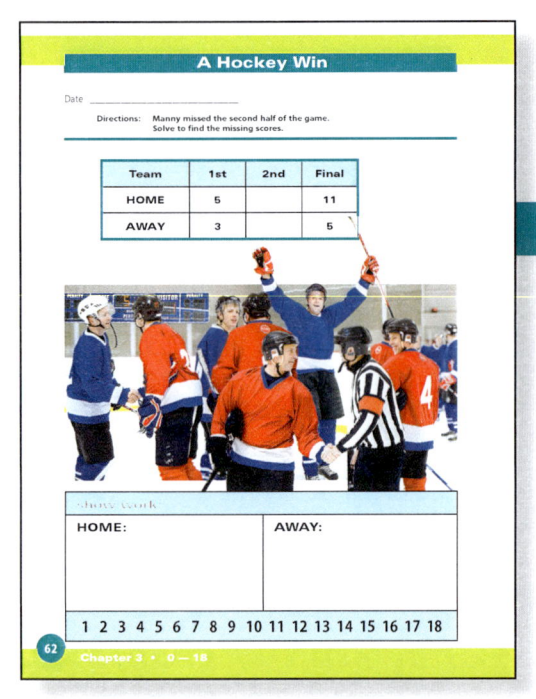

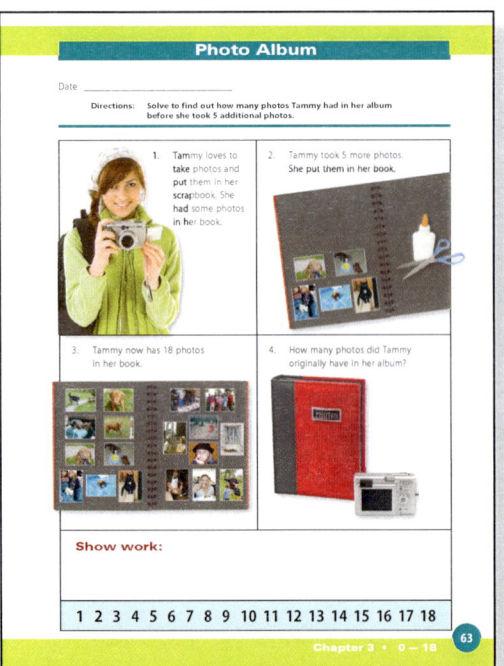

Materials

- teacher-made addition problems on the board, each with a missing addend
- number line
- dry board
- markers
- pencils
- student worksheets, pp. 62–63: A Hockey Win and Photo Album

Procedure

1. Write addition problems on the board, some with the first addend missing and some with the second addend missing, e.g.,
   8 + ___ = 12 and ___ + 6 = 14.

2. Read the problem as it is written.

3. Point to the box.

4. Tell students that the two numbers when added must equal the sum, or total amount.

5. Point to the number, e.g., 8. Point to the number line and count up 9, 10, 11, 12. Say: "We counted four numbers. The number 4 goes in the box."

6. Write 4 in the box. Say, "8 + 4 = 12."

7. Ask a student to check the problem to be sure that the answer is correct.

8. Continue until all of the problems have been solved.

Assign the worksheets:
A Hockey Win and **Photo Album**

Lesson 8

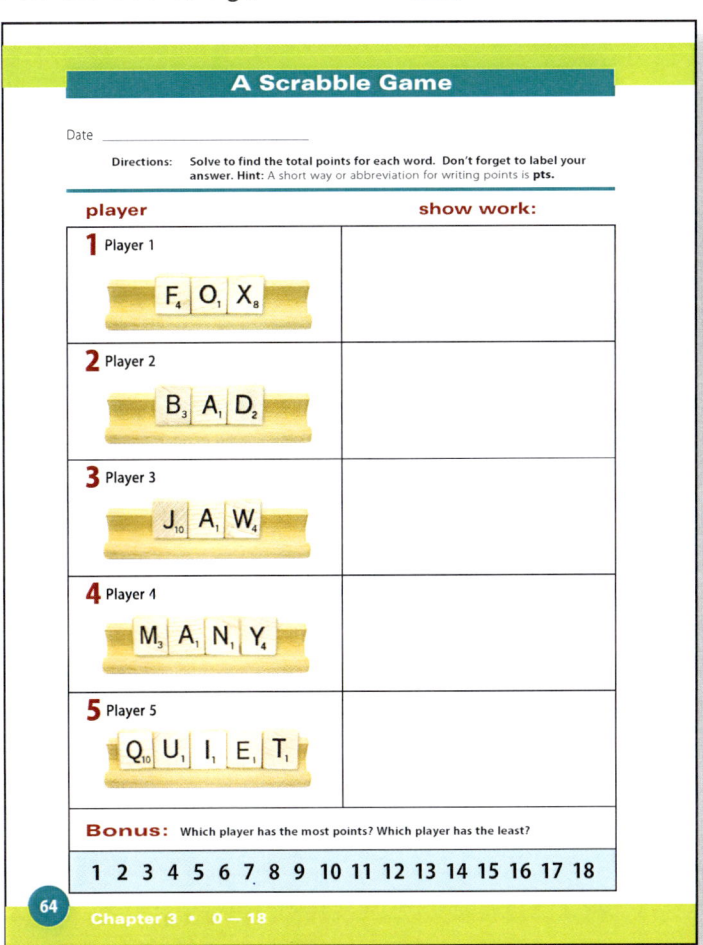

Objective

S. will solve a column addition problem up to eighteen.

Materials

- teacher-made single addition problems on the board, with sums to 18
- dry board
- markers
- scrabble pieces with 3-letter words
- pencils
- student worksheet, p. 64: **A Scrabble Game**

Procedure

1. Make a three-letter word using the scrabble pieces.
2. Point to the numbers on the pieces and say, "To find out how much this word is worth, we need to add the numbers that are on the pieces."
3. Write column addition problems on the board, e.g., 5 + 7 + 3 = ___.
4. Demonstrate how to solve the problem by adding the first two numbers together and then the final number to solve for the sum.
5. Give student volunteers opportunities to solve other problems.
6. Continue until all of the problems have been solved.
7. Assign words made from the scrabble pieces to individual students.
8. Students are to solve problems to find out how many points each word gets.
9. Continue until students are firm.

Assign the worksheet:
A Scrabble Game

Lesson 9

Objective

Using a map S. will solve to find the distance traveled on a short trip.

Materials

- teacher-made map of towns with miles between the different towns, or commercial maps of counties, etc.
- pencils
- student worksheet, p. 65: A Weekend Trip

Procedure

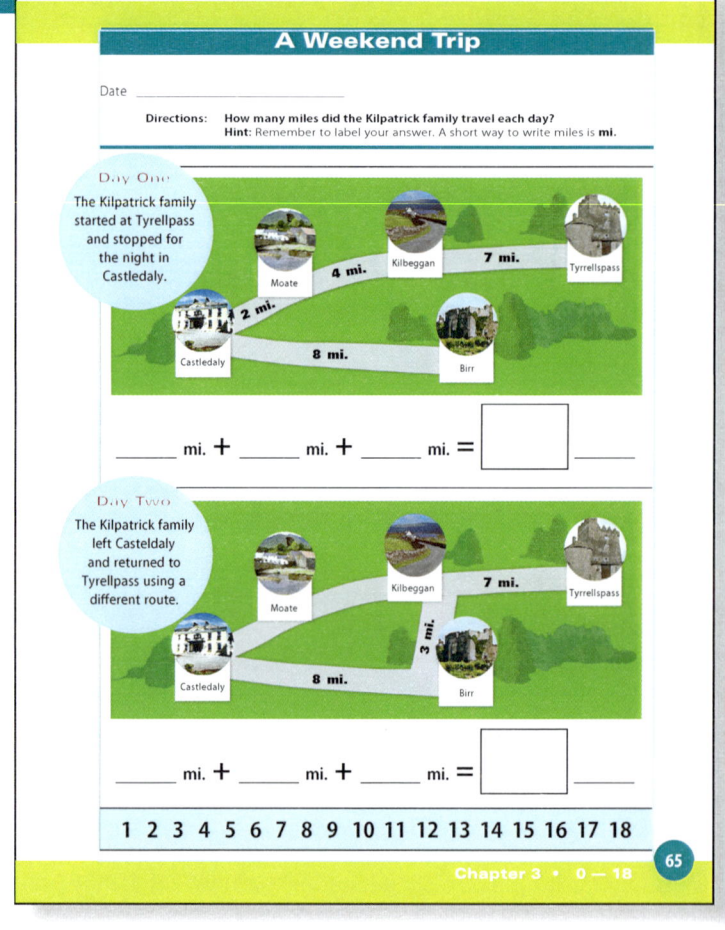

1. Show examples of maps and point out important features.
2. Explain how to read the miles between towns.
3. Point to the teacher-made map.
4. State that when traveling, especially by car, people use maps to find towns and cities on their trip.
5. They also use maps to determine the distances between 2 towns or cities.
6. Read the names of the towns on the map.
7. Point to the numbers between the towns and say that these numbers tell how many miles there are between towns.
8. Ask several student volunteers to read the miles between the towns.
9. Create scenarios about a family traveling to different towns.
10. Students must add the miles the "family" took to find out how far they traveled.
11. Increase the number of towns the family travels to 3, so that students must use column addition to solve for the total miles.
12. Continue until students are firm.

Assign the worksheet: **A Weekend Trip**

Lesson 10

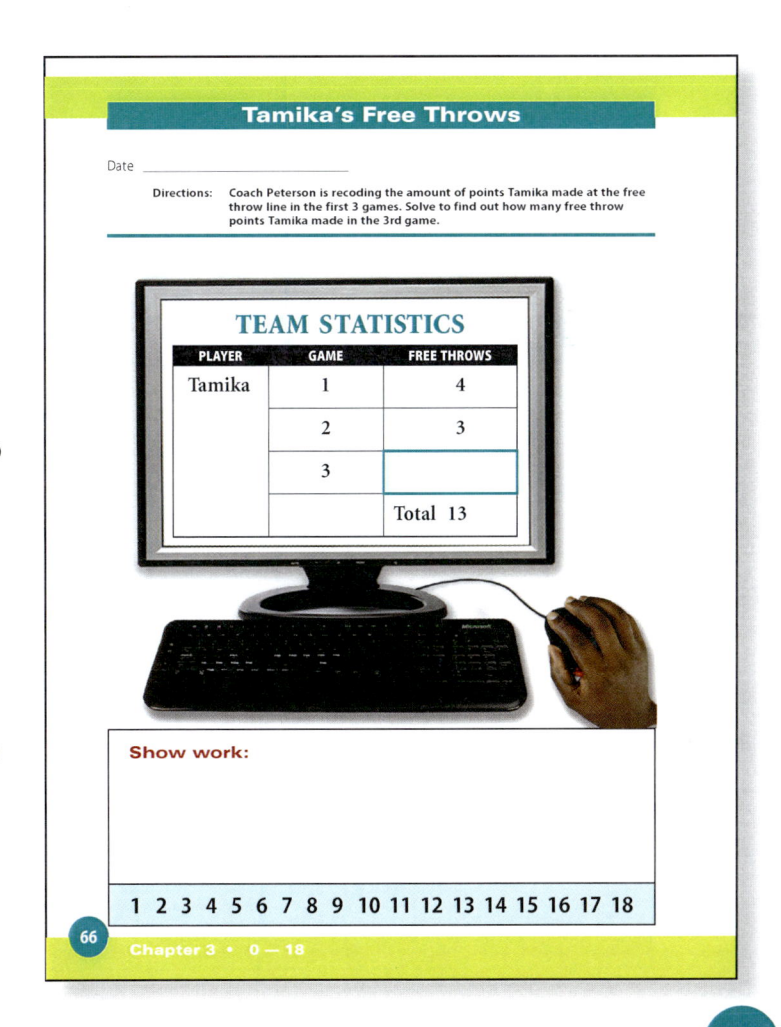

Objective

S. will solve a column addition problem with a missing addend.

Materials

- teacher-made column addition problems with a missing addend on the board
- number line
- dry board
- markers
- pencils
- student worksheet, p. 66: Tamika's Free Throws

Procedure

1. Review with students how to solving missing addends. (See Lesson 7.)
2. Teach students a similar strategy using column addition, e.g., 6 + 5 + ☐ = 16.
3. Read the problem as it is on the board.
4. Point to the box.
5. Tell students that the missing number that goes in the box must equal the sum, or total amount, when added to the other two numbers.
6. Point to 6 and add 5 to the 6 using the number line.
7. Say, "5 plus 6 equals 11. Now we have to find out how many more to add to get to 16."
8. Count up: "12, 13, 14, 15, 16."
9. Say: "We counted 5 numbers. The number 5 goes in the box."
10. Write 5 in the box.
11. Ask a student to check the problem using the number line.
12. Continue until all of the problems have been solved.

Assign the worksheet:
Tamika's Free Throws

Lesson 11

Objective

S. will solve a subtraction word problem to eighteen.

Materials

- small box
- set of small manipulatives
- pencils
- student worksheet, p. 67: Stocking Shelves

Procedure

1. Write the word **left** on the board.

2. Tell students that when this word appears in a problem, they will subtract.

3. Count a set of manipulatives, e.g., 14. Point to the small empty box. Tell students that some of the manipulatives will be put into the box. Count out a set, e.g., "1, 2, 3, 4, 5, 6, 7, 8." Tell students that there were 14 _____ and 8 are in the box. Ask: "How many are left in my hand?"

4. Demonstrate how to solve the problem. Write 14 − 8 = ___ on the board. Count the number of manipulatives that are left. Write 6 in the box and say the problem.

5. Repeat, using another problem.

6. Continue until students are firm.

Assign the worksheet: **Stocking Shelves**

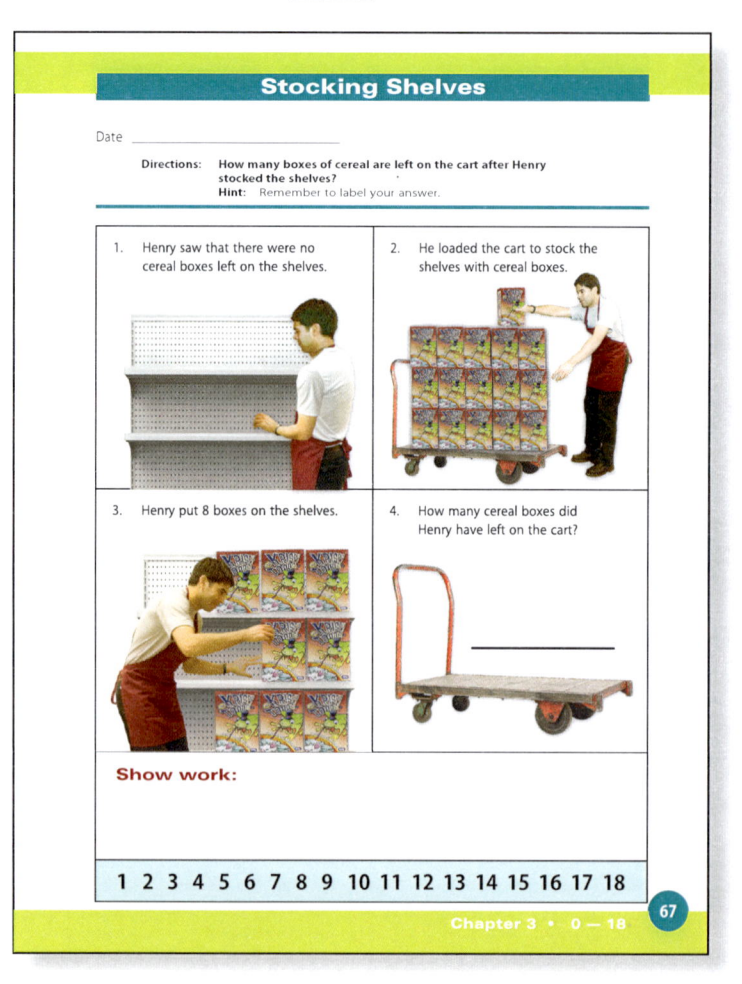

Lesson 12

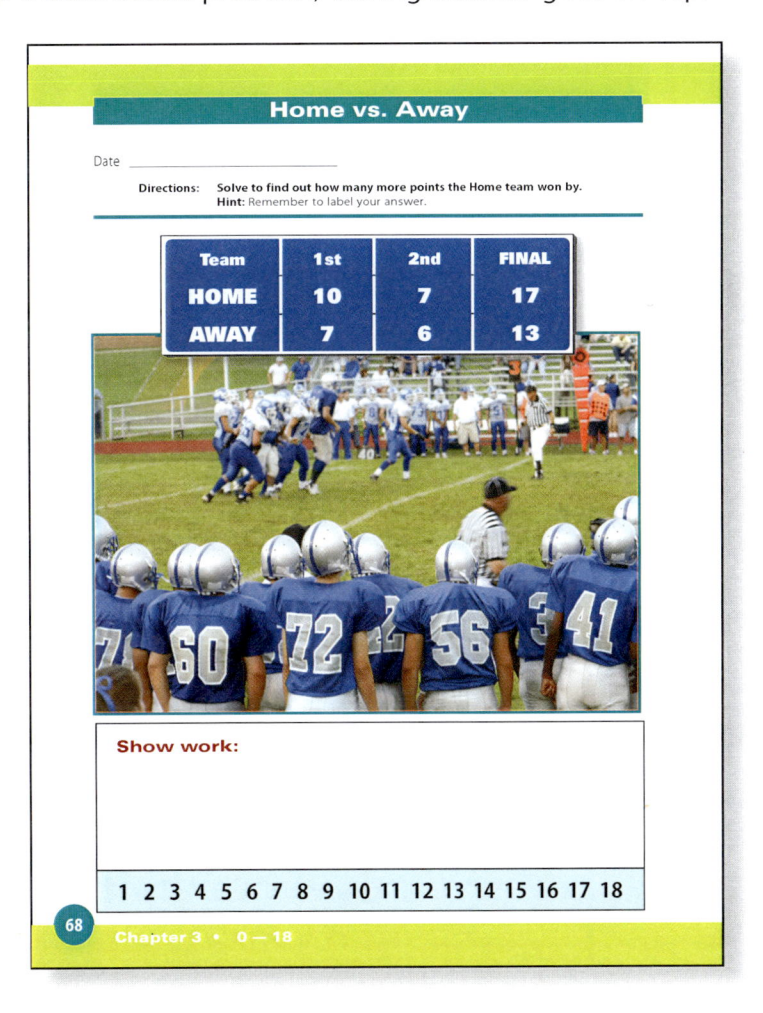

Objective

S. will solve a subtraction word problem to eighteen.

Materials

- teacher-made football scoreboard on the board
- pencils
- student worksheet, p. 68: Home vs. Away

Procedure

1. Write a score on the board, one for each team (see worksheet for an example).

2. Tell the students that the game is over. The scoreboard shows the final score.

3. The students must solve to find out how many more points the winning team had than the losing team.

4. Reinforce that when setting up a subtraction problem, the big number goes on top.

5. Point to the big number.

6. Write the number on the board, then subtract the smaller number from it to find out how many more points the winning team had.

7. Reinforce the concept that word problems must have a label. In this case the label is "points."

8. Continue using different scores and have student volunteers solve the problems.

Assign the worksheet:
Home vs. Away

Lesson 13

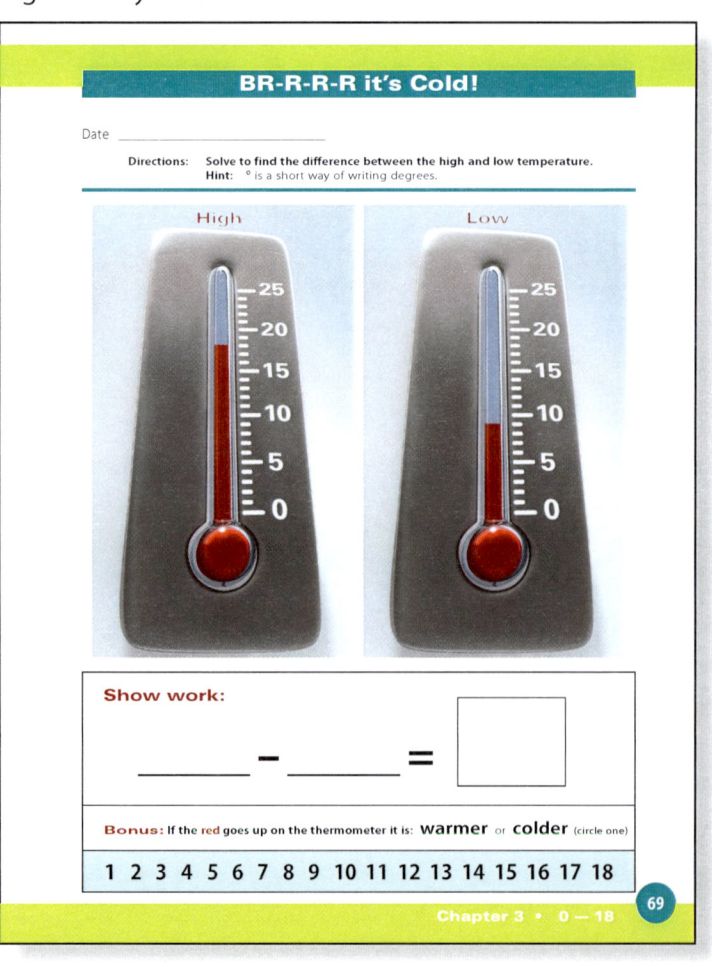

Objective

S. will find the difference between the high and low daily temperatures.

Materials

- 2 teacher-made thermometers on the board; label one "high" and the other "low"
- dry board
- markers
- pencils
- student worksheet, p. 69: BR-R-R-R It's Cold

Procedure

1. Point to the thermometer and review how to read the temperatures on an outdoor thermometer.

2. Review how to write degrees using the ° symbol.

3. Fill in a temperature on the thermometer labeled "high" and then the one labeled "low."

4. Subtract the low temperature from the high temperature to find the difference.

5. Remember to reinforce "degrees" as the label.

6. Continue with several more examples until the students can subtract the high and low temperatures without prompts.

Assign the worksheet:
BR-R-R-R It's Cold

Lesson 14

Objective

S. will find the difference between the high and low daily temperatures.

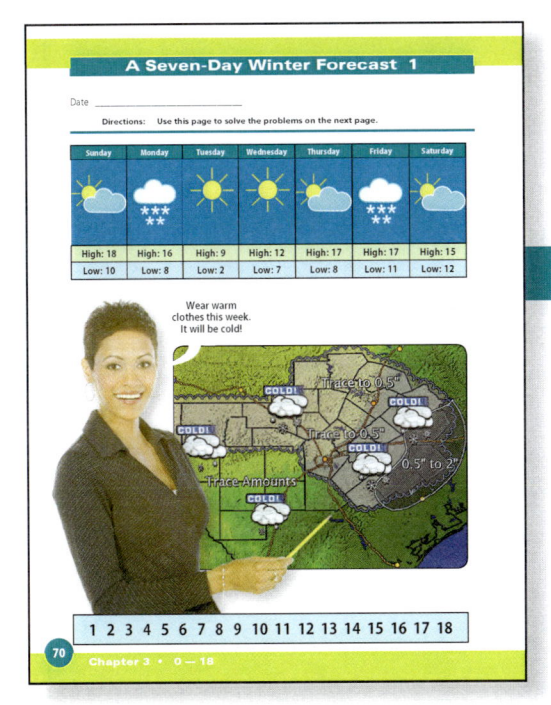

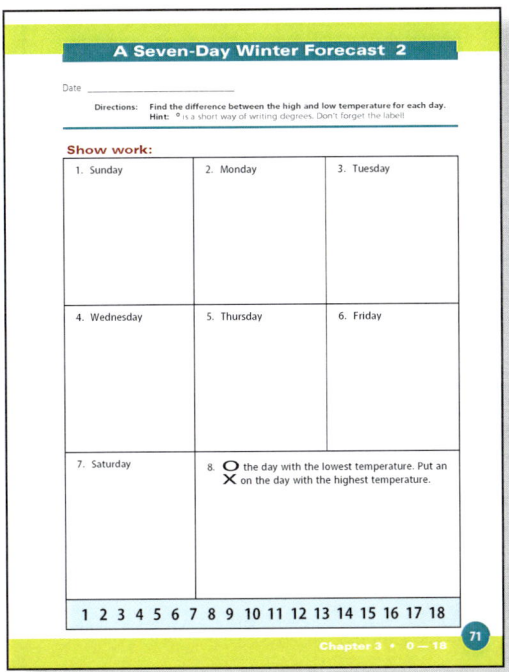

Materials

- examples of weeklong forecasts from newspapers
- teacher-made weeklong forecast of daily highs and lows
- dry board
- markers
- pencils
- student worksheets, pp. 70–71: A Seven-Day Winter Forecast 1 and 2

Procedure

1. Show examples of forecasts from the newspapers.
2. Point out important features.
3. Demonstrate how to read the highs and lows.
4. Point to the weeklong forecast on the board and call on individual students to read the temperatures for each day.
5. Pick a day to review how to find the difference between the high and low temperatures in the forecast.
6. Lead the students through the task.
7. Give student volunteers an opportunity to solve for the differences until all the days have been used.

Assign the worksheets:
A Seven-Day Winter Forecast 1 and 2

Lesson 15

Objective

S. will solve a subtraction word problem to eighteen.

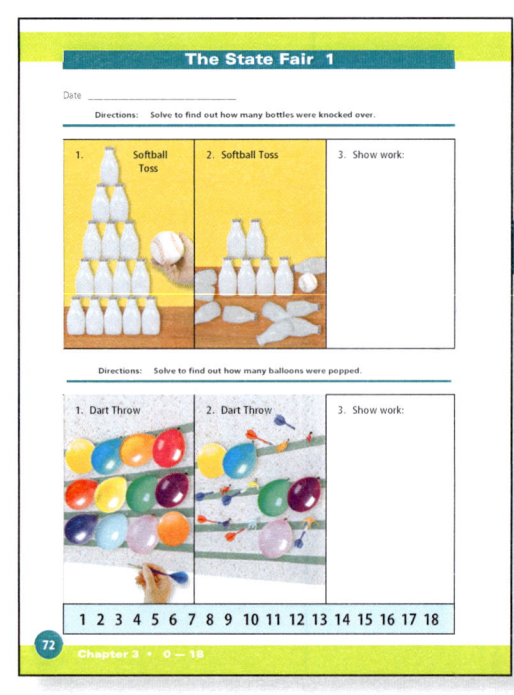

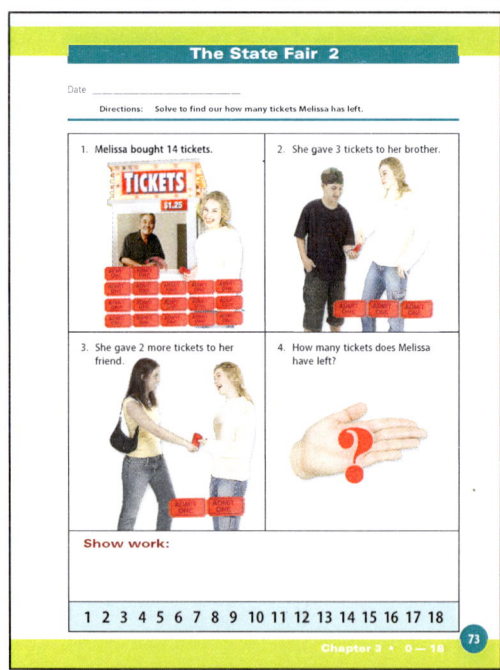

Materials

- milk cartons
- soft rubber ball
- teacher-made tickets
- dry board
- markers
- pencils
- student worksheets, pp. 72–73: The State Fair 1 and 2

Procedure

1. Elicit from students a list of carnival games one can find at a state fair, e.g., softball toss.

2. Stack the milk cartons and let students knock them down, keeping score of how many were knocked down.

3. Write the problem on the board for each throw. For example: There were 12 bottles. Henry knocked 7 of them down. How many were left standing?

4. Ask individual students to solve the problems.

5. Continue until all students have had an opportunity to solve a problem.

6. Create problems using the tickets and follow the same procedure.

Assign the worksheets:
The State Fair 1 and **2**

Lesson 16

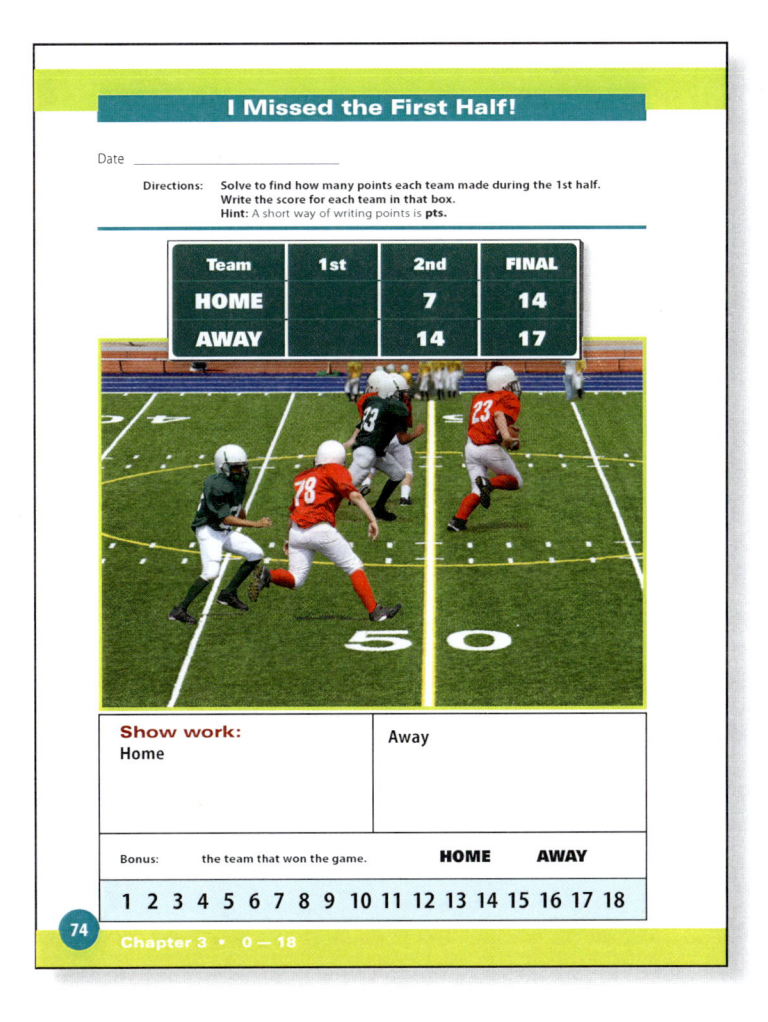

Objective

S. will solve to find a missing minuend.

Materials

- teacher-made scoreboard (see worksheet for an example)
- dry board
- markers
- pencils
- student worksheet, p. 74: I Missed the First Half!

Procedure

1. Draw the scoreboard on the dry board.

2. Put in a score for the second half and a final score for each team.

3. Tell students that they need to find out what the score was for the first half.

4. Write the final score on the board. Subtract the score for the second half from the final score.

5. The remainder is how many points a team scored in the first half.

6. Follow the same procedure for the second team.

7. Continue with additional scores until the students can complete the problem without prompts.

Assign the worksheet:
I Missed the First Half!

Lesson 17

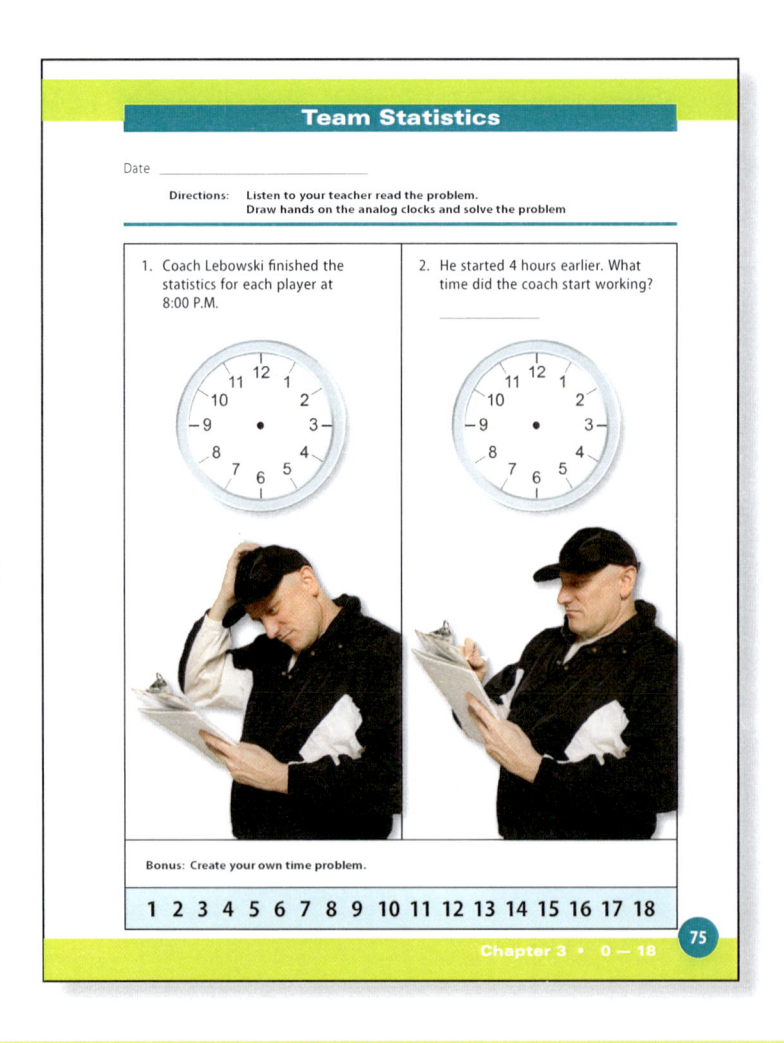

Objective

S. will draw hands on an analog clock to show a time that is more than one hour earlier from a set time.

Materials

- analog clock(s)
- teacher-made list of digital times to the hour
- dry board,
- markers
- pencils
- student worksheet, p. 75: Team Statistics

Procedure

1. Pointing to a digital time on the teacher-made list, review setting the analog clock to the digital time.

2. Continue until students are firm.

3. Next, demonstrate how to move the hands on the analog clock so that it is an hour earlier, by turning the hands counterclockwise and counting by 5s around the clock.

4. Lead students through the task.

5. Continue by having students set an analog clock 2 or more hours earlier.

Assign the worksheet:
Team Statistics

Lesson 18

Objective

S. will draw hands on an analog clock to show a time one-half hour later than a set time.

Materials

- analog clock(s)
- teacher-made list of digital times
- dry board
- markers
- pencils
- student worksheet, p. 76: Walking the Dog

Procedure

1. Review telling time to the hour using an analog clock with the students.
2. Ask student volunteers to find times to the hour that are earlier and later than a stated time.
3. Point to a digital time on the teacher list.
4. Next, counting by five, move the hands on the analog clock to indicate a time that is a half hour later.
5. Lead students through the task.
6. Create scenarios where students must complete a task which takes a half hour to do.
7. Students take turns setting the analog clock until all teacher-made times have been used.

Assign the worksheet:
Walking the Dog

Note: Use the Hundreds Chart in Chapter 3 for students who need to practice counting by 5s. Make a clock on the board with 5 minutes at each number.

Lesson 19

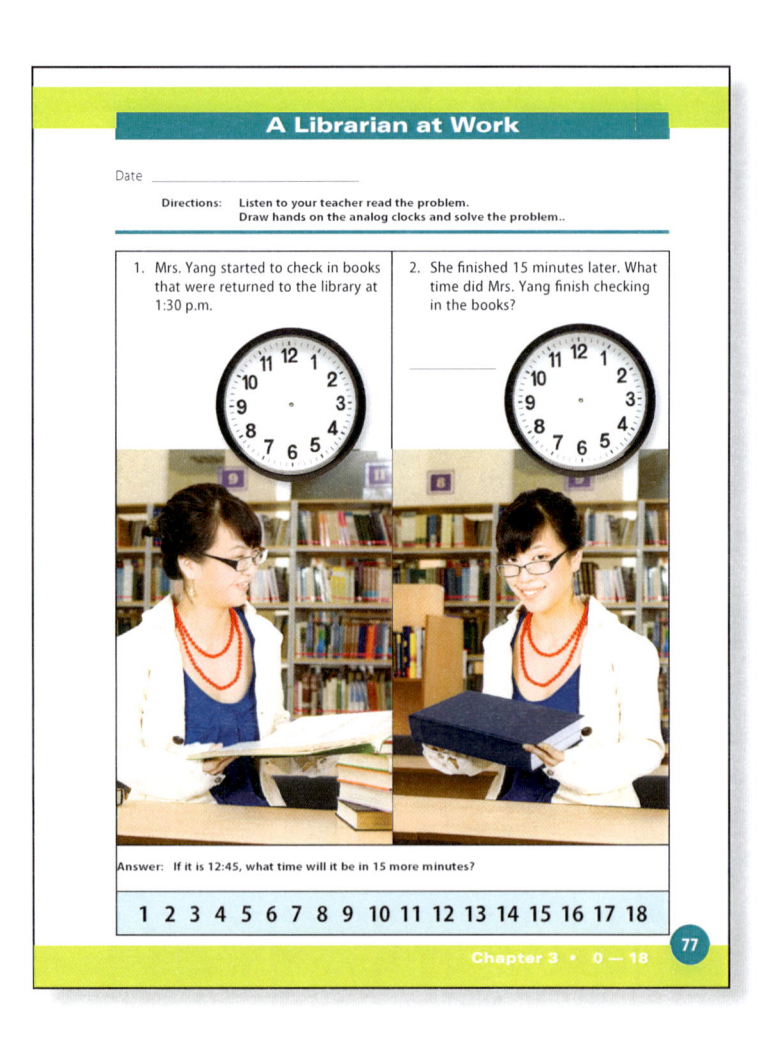

Objective

S. will draw hands on an analog clock that is a quarter hour later than a set time.

Materials

- analog clock(s)
- teacher-made list of digital times
- dry board
- markers
- pencils
- student worksheet, p. 77: A Librarian at Work

Procedure

1. Review with the students telling time to the hour and half hour.

2. Ask student volunteers to set the clock an hour or half hour later than a set time.

3. Point to a digital time on the board.

4. Counting by 5, demonstrate how to move the hands on the analog clock to indicate a time that is a quarter hour or 15 minutes later.

5. Lead students through the task.

6. Create scenarios where students must complete a task which takes 15 minutes to do.

7. Students take turns setting the analog clock until all teacher-made times have been used.

Assign the worksheet:
A Librarian at Work

A Librarian at Work

Date _____

Directions: Listen to your teacher read the problem. Draw hands on the analog clocks and solve the problem..

1. Mrs. Yang started to check in books that were returned to the library at 1:30 p.m.

2. She finished 15 minutes later. What time did Mrs. Yang finish checking in the books?

Answer: If it is 12:45, what time will it be in 15 more minutes?

1 2 3 4 5 6 7 8 9 10 11 12 13 14 15 16 17 18

Chapter 3 • 0 — 18

77

62

Explore Math Teacher's Manual

Lesson 20

Objective

S. will solve to find out earnings made during a set time.

Materials

- analog clock(s)
- teacher-made list of digital times
- 18 one-dollar bills
- dry board
- markers
- pencils
- student worksheets, pp. 78–79: A Weekend Job 1 and 2

Procedure

1. Tell students that one reason for learning elapsed time is that they need to figure out how much money they will earn in an hour, two hours, etc., while on a job.
2. Point to a digital time review, telling time to the hour, half hour, and quarter hour.
3. Create scenarios where students must work for at least an hour and determine the total pay for each job, e.g., Sara earns $5.00 an hour baby-sitting for a family. She worked two hours. How much money did Sara earn?
4. Set the analog clock to a time.
5. Move the hands around an hour and say, "Sara worked one hour."
6. Lay out five dollars.
7. Move the hand around the clock again and say, "Sara worked a second hour."
8. Lay out another five dollars.
9. Ask a student volunteer to count the amount of money that Sara earned.
10. Continue with other scenarios until all the students have had an opportunity to solve a problem.
11. Be sure to include hours and half hours in the scenarios.
12. Give students analog clocks to solve the problems on the worksheet.

Assign worksheets: **A Weekend Job 1** and **2**

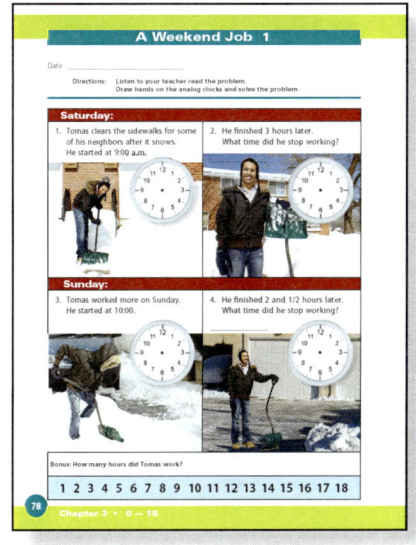

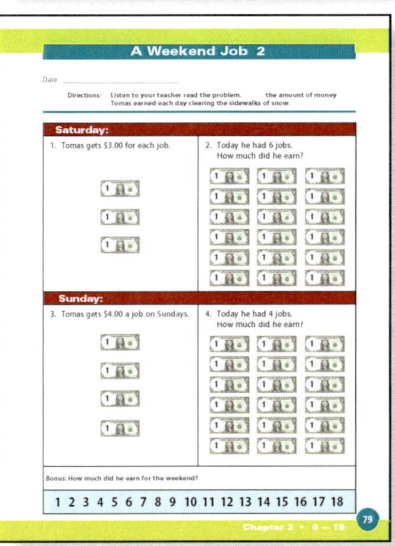

Lesson 21

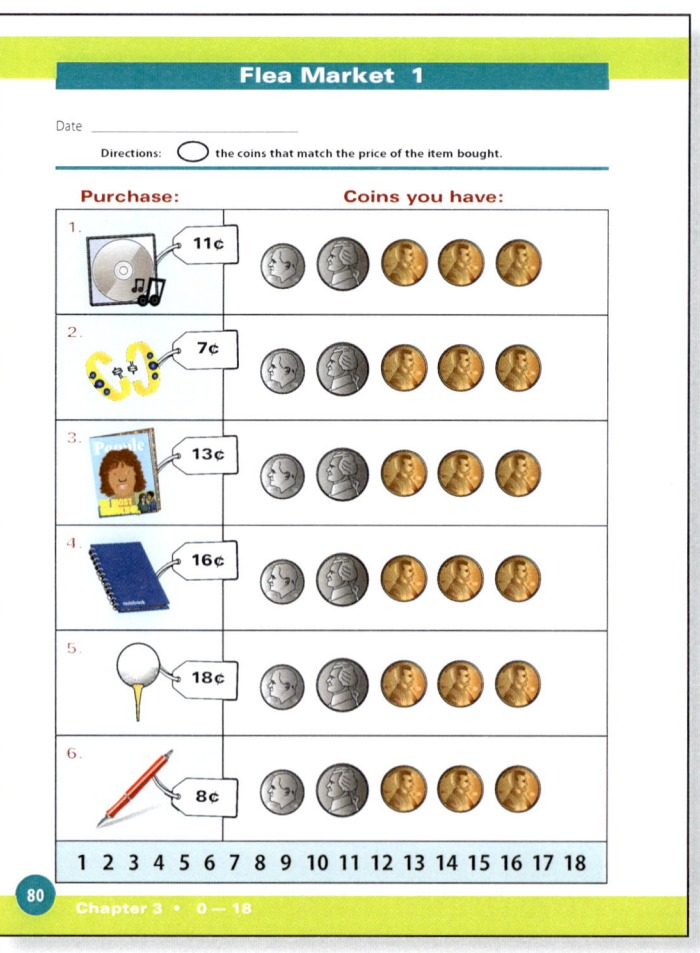

Objective

S. will match the purchase price of an item using dimes, nickels, and pennies to eighteen cents.

Materials

- small items to purchase with price tags to 18 cents
- sets of nickels, dimes, and pennies to 18 cents
- pencils
- student worksheet, p. 80: Flea Market 1

Procedure

1. Place the items to purchase in a row in front of the students.

2. Select an item and a set of coins.

3. Count out the coins to match the price of the item. The item cannot cost more than the money in the set.

4. Use the smallest number of coins possible.

5. Repeat the procedure with another item.

6. Give each student a set of coins. One at a time, the students make a purchase and count out the coins to match the price of the item.

7. Continue until all of the students have an opportunity to make a purchase.

Assign the worksheet:
Flea Market 1

Lesson 22

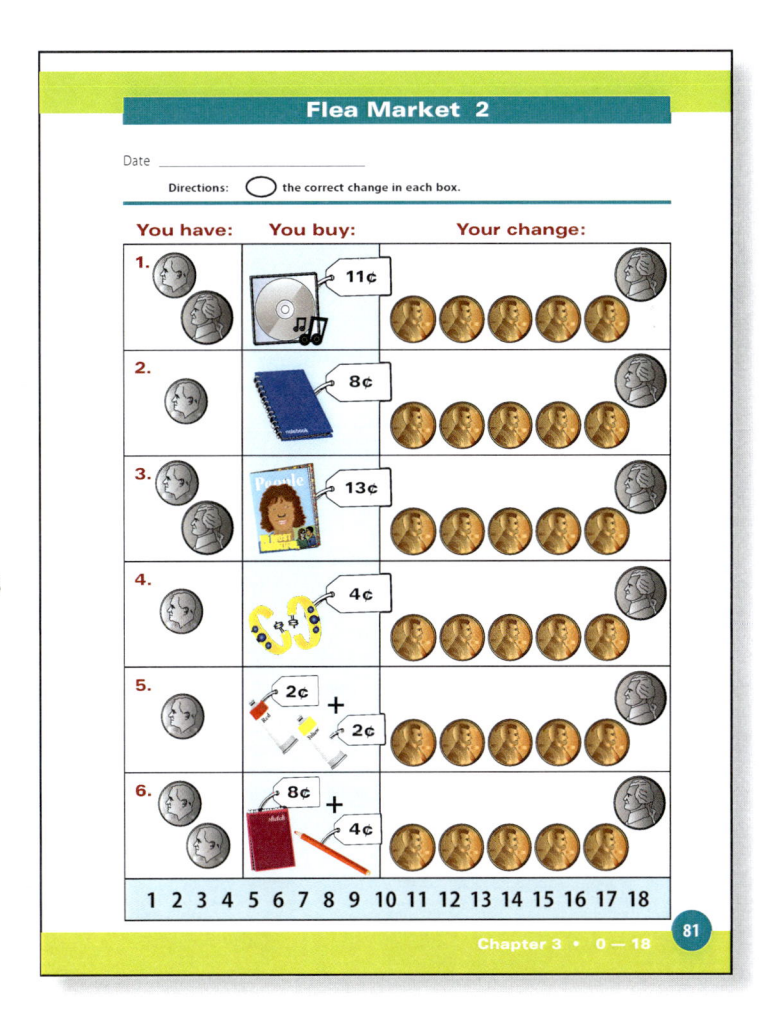

Objective

S. will count change back from eighteen cents.

Materials

- small items to purchase with price tags to 18 cents
- sets of dimes, nickels and pennies
- pencils
- student worksheet, p. 81: Flea Market 2

Procedure

1. Place items to purchase in a row in front of the students,
2. Count out a set of money and use the money to purchase an item.
3. The item cannot cost more than the money in the set.
4. Give the clerk the money.
5. The clerk counts back the change.
6. Lead students through the task.
7. Assign students pairs; one is the customer and the other the clerk. The customer buys an item and the clerk counts back the change.
8. Students switch roles, continuing to purchase items and count back change until they are firm.

Assign the worksheet:
Flea Market 2

Lesson 23

Objective

S. will match the purchase price of an item using dimes, nickels, and pennies to eighteen cents.

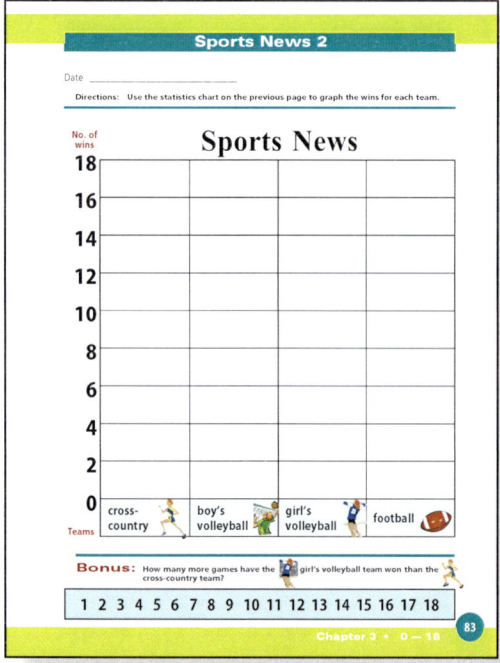

Materials

- small items to purchase with price tags to 18 cents
- sets of nickels, dimes, and pennies to 18 cents
- pencils
- student worksheets, pp. 82–83: Sports News 1 and 2

Procedure

1. Find the win/loss column of a sport for the regional high school games.

2. Write the names of each team and wins so far this season.

3. Make a bar graph on the board.

4. Review the parts of the graph with the students.

5. Tell the students that they will graph the wins for each team.

6. Demonstrate how to fill in the first columns of the graph.

7. Call on individual students to complete the graph.

8. When the graph is finished, ask questions based upon the data, e.g., what team had the most wins?

Assign the worksheets: **Sports News 1** and **2**

Lesson 24

Objective

S. will plot and interpret information from a bar graph.

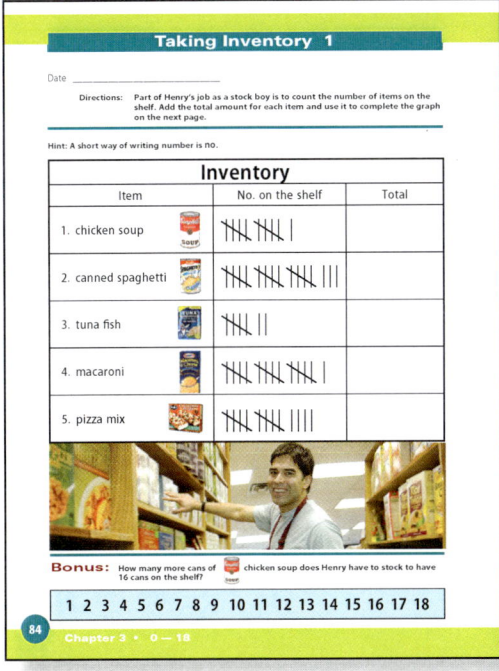

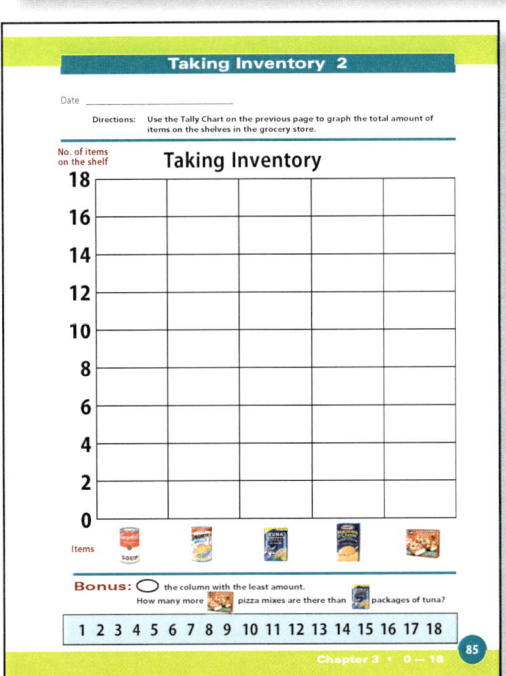

Materials

- teacher-made tally chart and graph
- dry board
- markers
- pencils
- student worksheets, pp. 84–85: Taking Inventory 1 and 2

Procedure

1. Tell students that on some jobs people need to take inventory.

2. Inventory is merchandise or stock that a store or another business has on hand to sell.

3. Brainstorm jobs that may require people to take inventory.

4. Create a scenario where a person must take inventory, e.g., a stock person in a grocery store. On the tally sheet, list five items that are sold on grocery store shelves.

5. Make tally marks for each item on a shelf. Vary the tally marks for the items.

6. Demonstrate how to total the tally marks and write numbers to match the tallies.

7. Review the parts of the graph with the students.

8. Use the tallies from the tally chart to plot onto the graph.

9. Demonstrate how to complete the first column.

10. Call on student volunteers to complete the next columns.

11. Ask questions about the graph when it is finished.

Assign the worksheets: **Taking Inventory 1** and **2**

67

Chapter 3 • 0–18

Lesson 25

Objective

S. will skip count by two or three to solve a problem.

Materials

- teacher-created problems
- pencils
- student worksheet, p. 86: Cafeteria Orders

Procedure

1. Create problems that require skip counting by using the students in the room, e.g., select three students. Give each student two pencils. Tell the group that each student has two pencils. How many pencils will they have altogether?

2. Ask a student to solve and check the answer.

3. Continue until all students have had an opportunity to solve a problem using skip counting.

Assign the worksheet:
Cafeteria Orders

Note: For students who have difficulty skip counting, use the Hundreds Chart in Chapter 3.

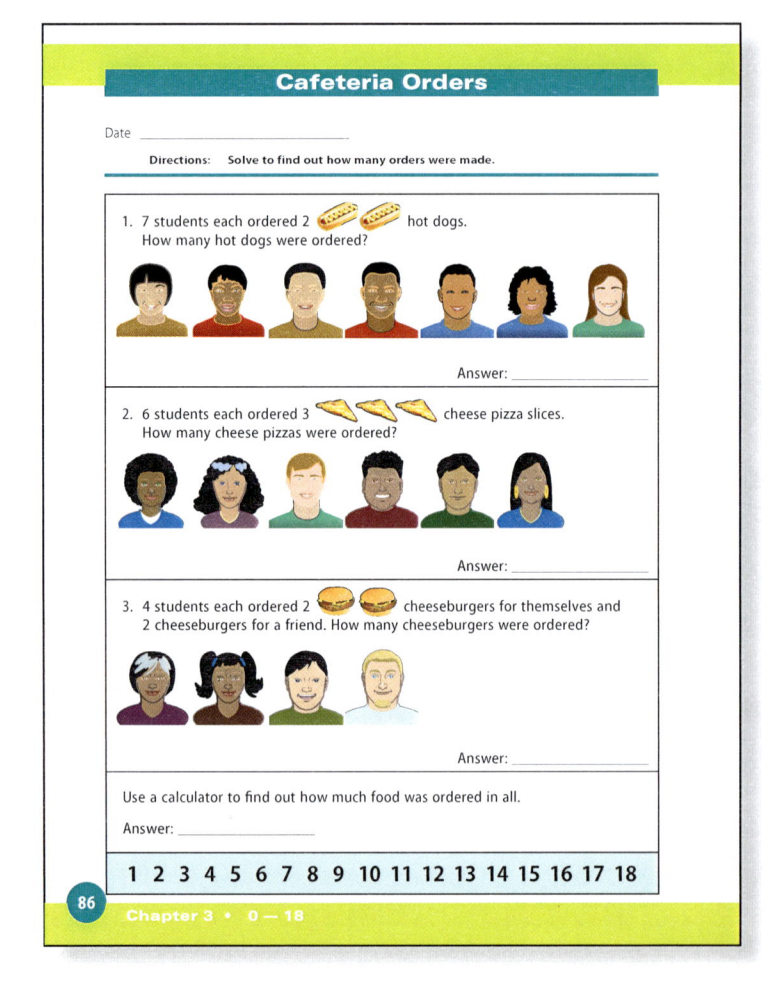

Cafeteria Orders

Date _____

Directions: Solve to find out how many orders were made.

1. 7 students each ordered 2 hot dogs. How many hot dogs were ordered?

   Answer: _____

2. 6 students each ordered 3 cheese pizza slices. How many cheese pizzas were ordered?

   Answer: _____

3. 4 students each ordered 2 cheeseburgers for themselves and 2 cheeseburgers for a friend. How many cheeseburgers were ordered?

   Answer: _____

Use a calculator to find out how much food was ordered in all.
Answer: _____

1 2 3 4 5 6 7 8 9 10 11 12 13 14 15 16 17 18

Lesson 26

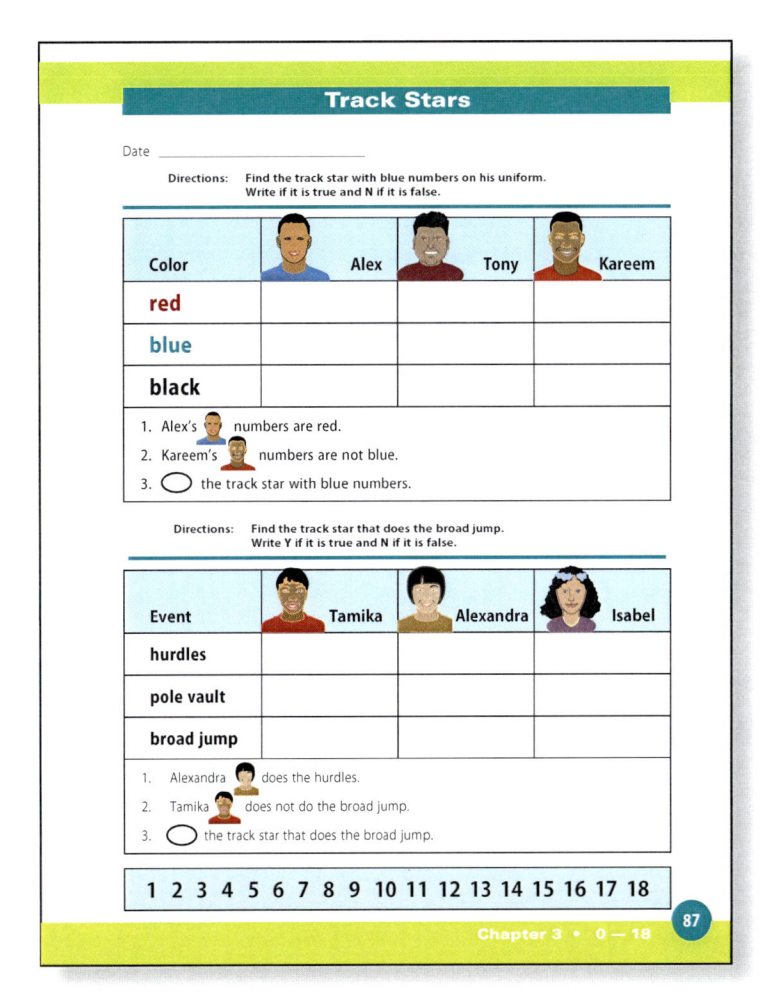

Objective

S. will read and follow clues to solve a math riddle.

Materials

- a teacher-made word problem
- pencils
- student worksheet, p. 87: Track Stars

Procedure

1. Create a word problem similar to the ones found on the Track Stars worksheet. For example, find the student who is wearing a striped shirt.

2. When saying the problem, point out important words to notice, such as "are" and "not".

3. Lead the students through the problem again.

4. Continue with new problems until the students are firm.

Assign the worksheet:
Track Stars

Chapter 4
0-100

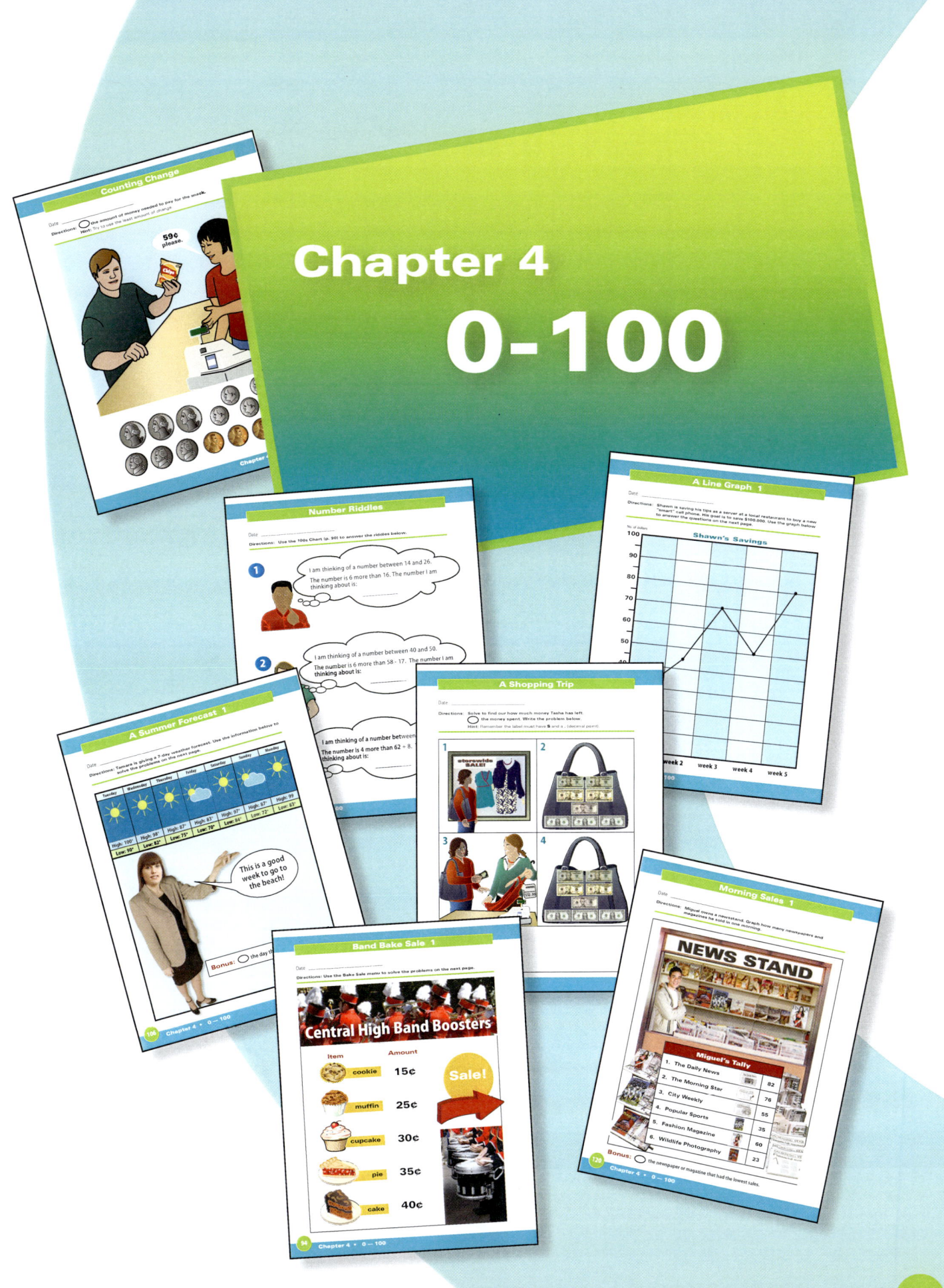

Chapter 4 • 0–100

Lesson 1

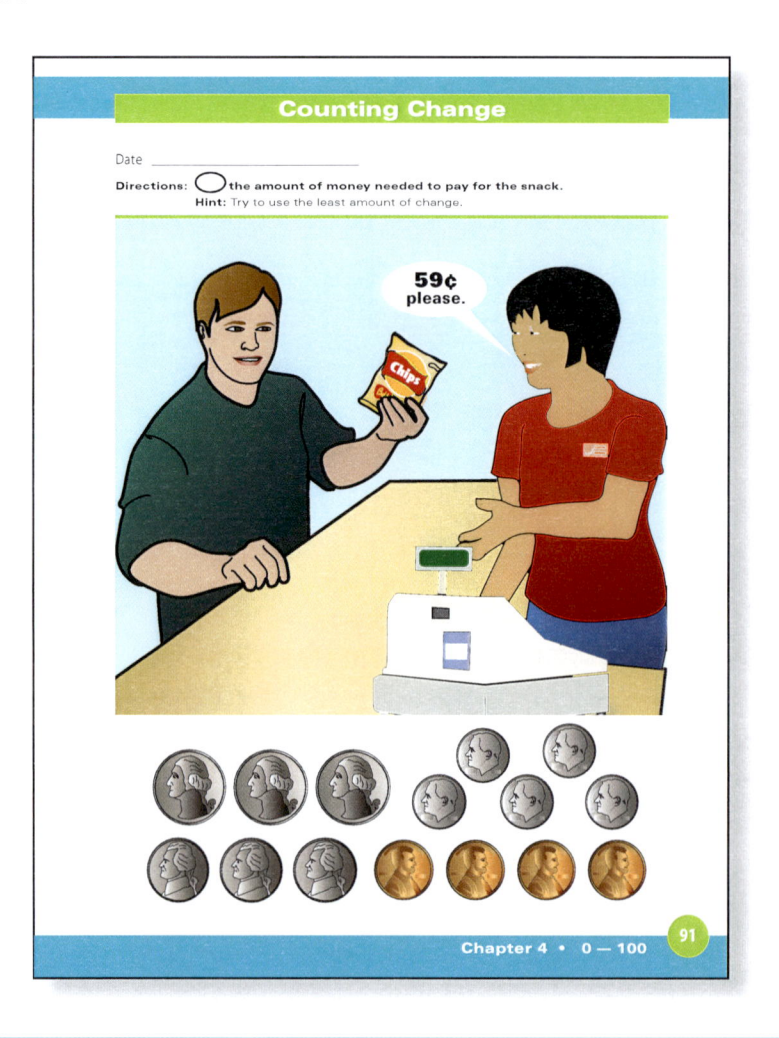

Objective

Using coins S. will match the total amount of money to the purchase price.

Materials

- sets of coins to a dollar
- variety of items with price tags on them
- pencils
- student worksheet, p. 91: Counting Change

Procedure

1. Create a scenario where students purchase items, e.g., variety store, grocery store, etc.//
2. Select an item to buy.
3. Count out the coins to match the purchase price.
4. Lead students through the task.
5. Assign students pairs, one to be a customer and the other the clerk.
6. The customer selects an item to buy.
7. The customer counts out the money to match the purchase price.
8. The clerk checks to be sure the amount is correct.
9. Students switch roles.
10. Continue until students are firm.

Assign the worksheet:
Counting Change

Explore Math Teacher's Manual

Lesson 2

Objective

Using a map S. will count how many blocks there are from one location to another.

Materials

- teacher-made map of a neighborhood or city with key
- city maps of the city where the students live
- dry board
- markers
- pencils
- student worksheet, p. 92: Center City

Procedure

1. Show examples of the city map.

2. Point out special features or places on the map.

3. Point to the teacher-made map. Tell students that this is a map of _____, and like the map of the city, has places and streets.

4. Explain the map key.

5. Encourage individual students to use the key for the map to identify places and streets.

6. Point to two places on the map. Demonstrate how to count the blocks from one place to another.

7. Lead students through the task.

8. Ask individual students to count blocks going from one place to another.

9. Continue until students are firm.

Assign the worksheet: **Center City**

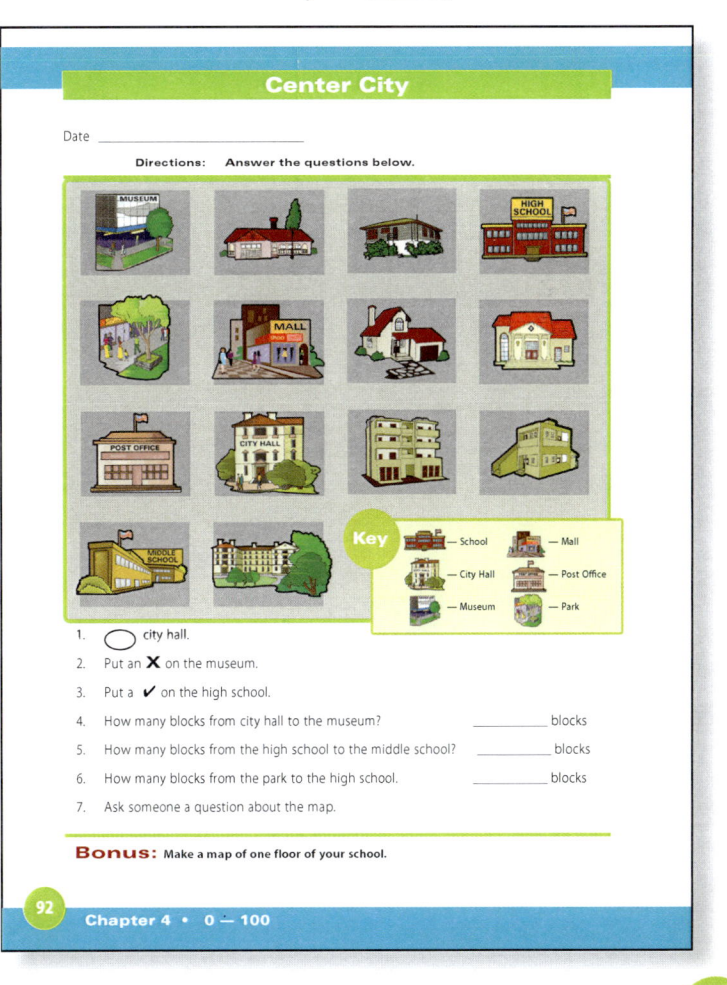

Lesson 3

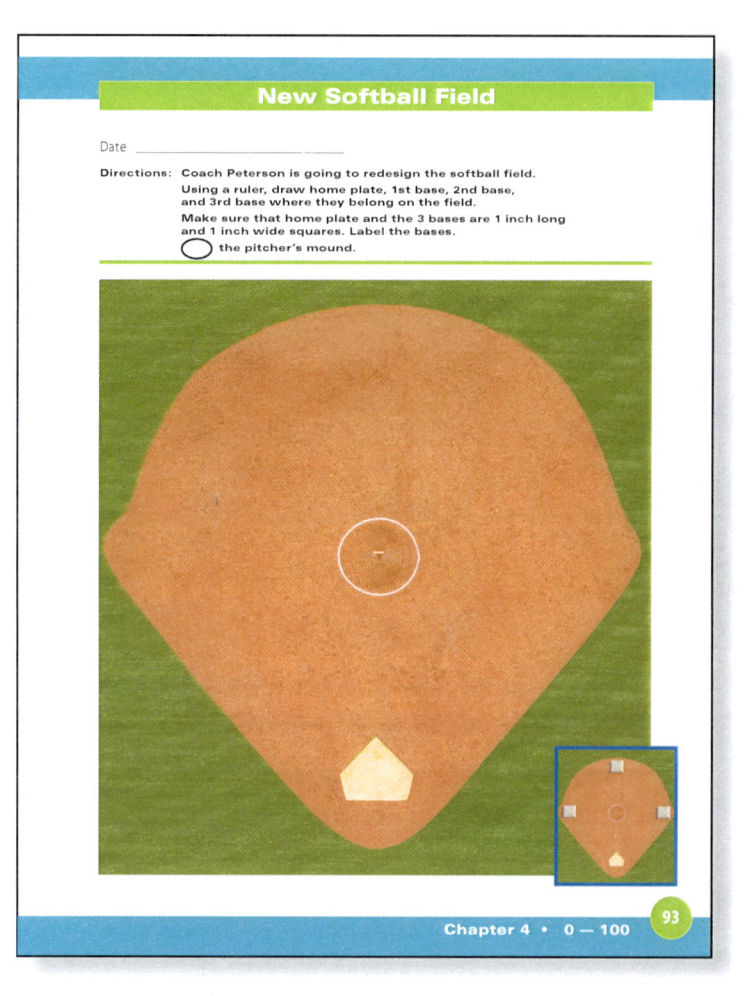

Objective

S. will measure and draw a square inch.

Materials

- rulers
- pencils
- student worksheet, p. 93: New Softball Field

Procedure

1. Explain to students that sometimes they will have to draw squares or other geometric shapes.

2. Brainstorm when people need to do this, e.g., buying a carpet, new furniture. Some workers use this skill, such as landscapers, architects etc.

3. Tell students that sometimes teachers need to do this when thinking about rearranging a classroom.

4. Draw an outline of a classroom on the board.

5. Tell students that there need to be ____ desks in the room and that each desk has to be one-inch by one-inch square.

6. Demonstrate on the board how to draw a square inch.

7. Lead students through the task.

8. Call on individual students to draw a square inch.

9. Continue until the students can do it without prompts.

Assign the worksheet:
New Softball Field

Lesson 4

Objective

Using a menu, S. will purchase items to one dollar.

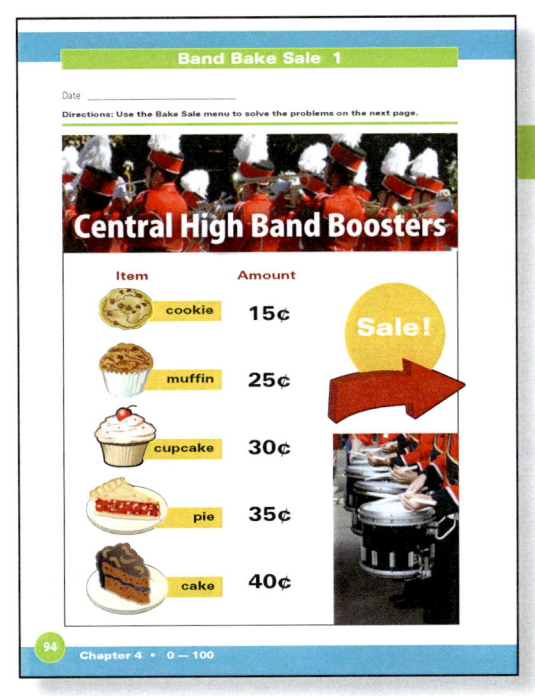

Materials

- real menus from restaurants
- teacher-made menu
- sets of coins
- pencils
- student worksheets, pp. 94-95: Band Bake Sale 1 and 2

Procedure

1. Show real menus.
2. Read the menus to the students.
3. Point to the teacher-made menu.
4. Read it and compare to real menus.
5. Select an item to buy on the teacher menu and count out the coins to match the purchase price.
6. Lead students through the task.
7. Select a set of coins and determine what item can be bought using the coins.
8. Lead the students through the task.
9. Repeat until firm.
10. Select a set of coins and determine how many items can be bought using the set.
11. Lead the students through the task.
12. Continue until students are firm.

Assign the worksheets: **Band Bake Sale 1** and **2**

Lesson 5

Objective

Using the counting on strategies S. will solve word problems.

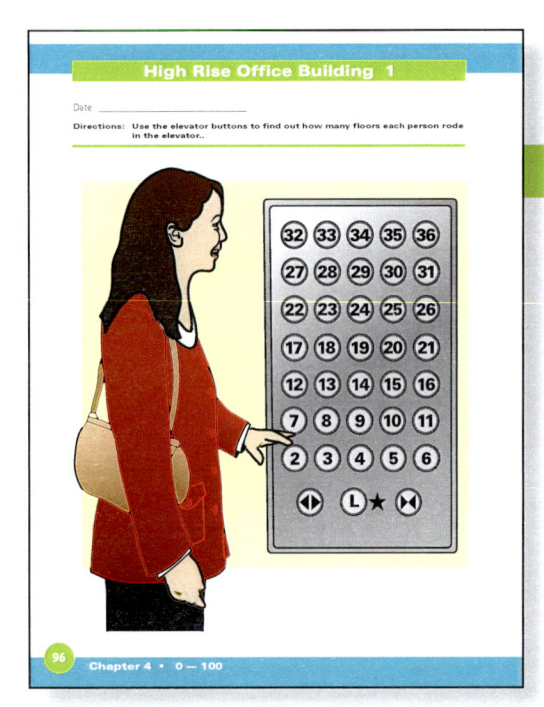

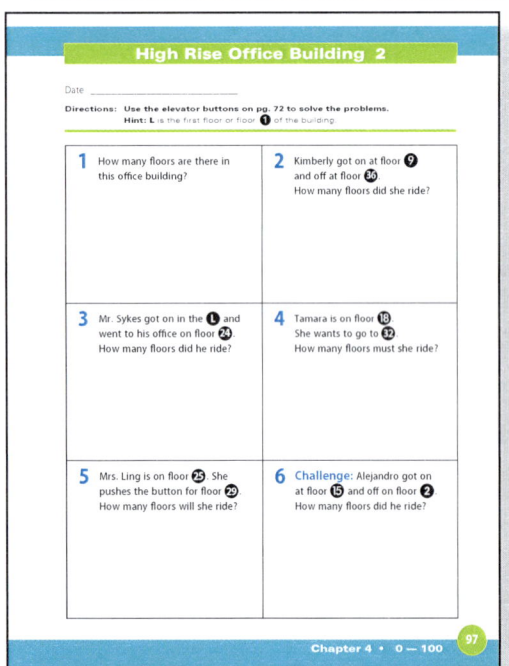

Materials

- teacher-made elevator panel on the board
- dry board
- markers
- pencils
- student worksheets, pp. 96-97: High-Rise Building 1 and 2

Procedure

1. Point to the elevator panel. Talk about the number of floors shown and what **L** means.

2. Create scenarios about passengers riding the elevator, e.g., Mrs., Gregory got on the elevator on floor 14 and got off on floor 22. How many floors did she ride? Using the elevator panel count up from 14 to 22 to find the answer, 8 floors.

3. Continue until students are firm.

Assign the worksheets:
High-rise Building 1 and **2**

Note: Some students may find that using a counting back strategy is easier.

Lesson 6

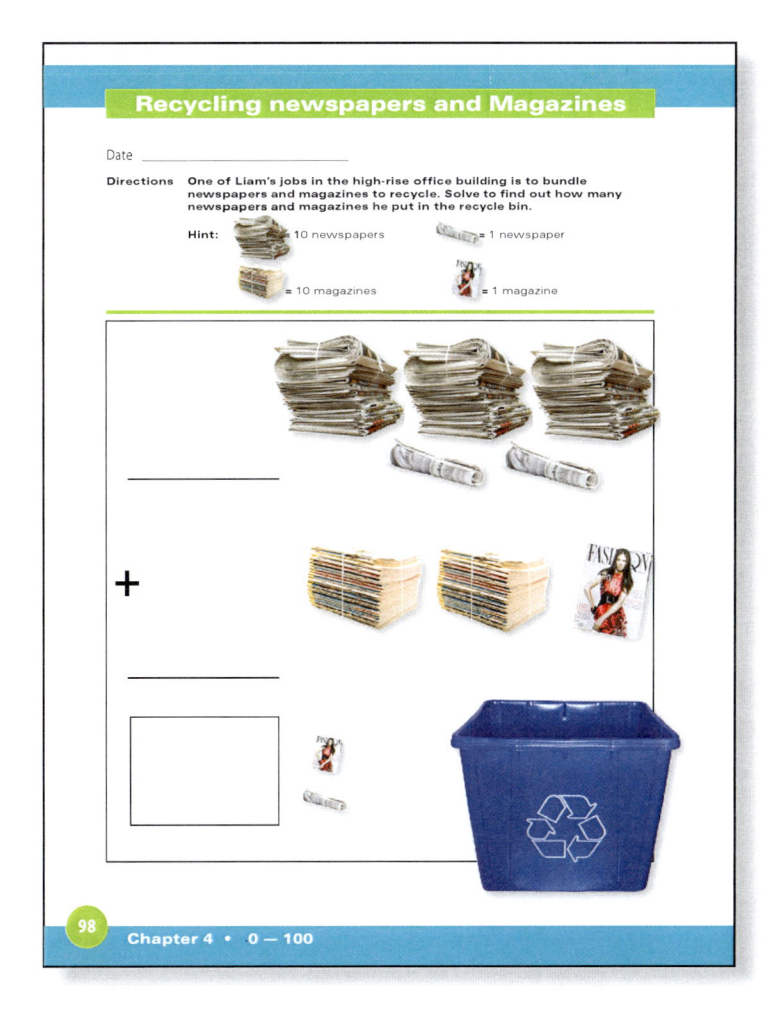

Objective

S. will solve an addition word problem to one hundred.

Materials

- counters: ones and tens
- pencils
- student worksheet, p. 98: Recycling Newspapers and Magazines

Procedure

1. Demonstrate how to use the counters.

2. Practice counting by tens, pointing to the ten rods as the students count.

3. Practice how to count by tens and then adding on ones.

4. Create several addition problems using the tens and ones for the students to solve.

5. Continue until the students are firm.

Assign worksheet:
Recycling Newspapers and Magazines

Lesson 7

Objective

S. will solve an addition word problem to one hundred

Materials

- counters, ones and tens
- pencils
- student worksheet, p. 99: Sandwiches for Sale.

Procedure

1. Review how to count by ten using the ten rods.

2. Review with the counters, counting by tens and adding on ones, e.g., 10, 20, 30, 31, 32, 33, 34.

3. Create several addition problems where the students must use the tens and ones counters to solve them.

4. Continue until students are firm.

Assign the worksheet:
Sandwiches for Sale

Lesson 8

Objective

S. will solve an addition word problem with a missing addend.

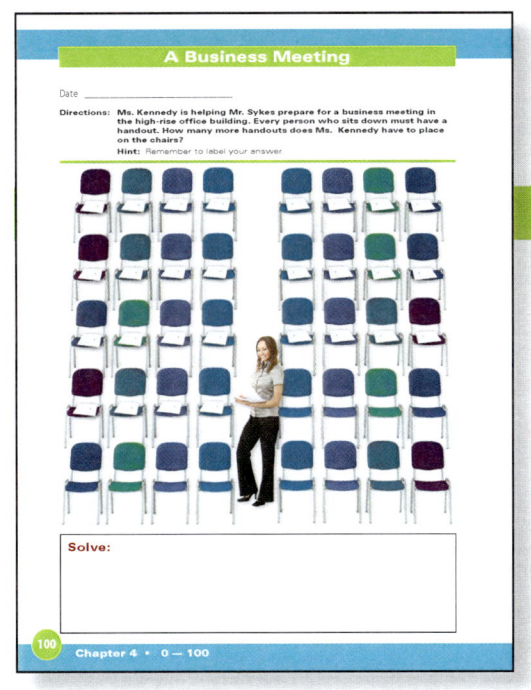

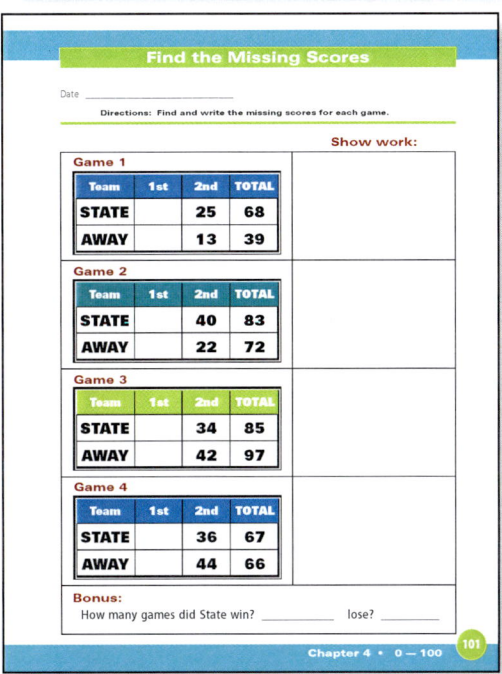

Materials

- teacher-made addition problems with one of the addends missing,
- dry board
- markers
- pencils
- student worksheets, pp. 100-101: A Business Meeting and Find the Missing Scores

Procedure

1. Write several addition problems on the board with a missing addend in each problem, e.g., 28 + _____ = 30 and _____ + 41 = 48.

2. Point to the box with the missing addend.

3. Create a scenario for the problem.

4. Demonstrate how to solve the problem.

5. Lead students through the process of solving the problem.

6. Continue until students can solve the problem without prompts.

Assign the worksheets:
A Business Meeting and **Find the Missing Scores**

Lesson 9

Objective

S. will solve a word problem using counting on strategies.

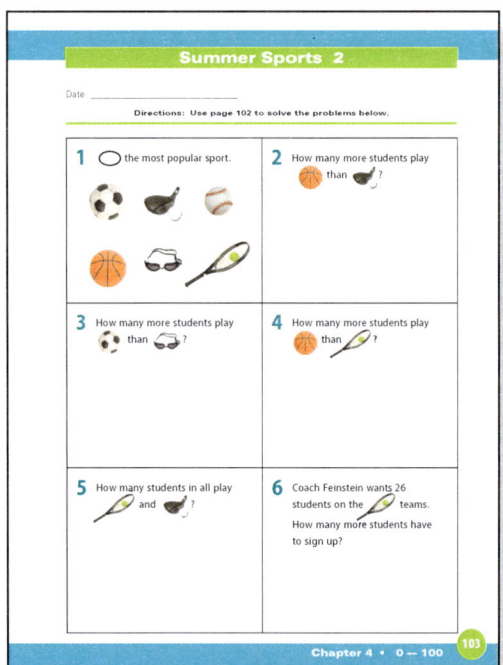

Materials

- a teacher-made list of items
- One Hundreds Chart from Student Workbook, p. 90
- dry board
- markers
- pencils
- student worksheets, pp. 102-103: Summer Sports 1 and 2

Procedure

1. Read the teacher-generated list of items.
2. Point to the list and ask students to tell what item has the most and what has the least.
3. Select two items and ask the students which item has more.
4. Tell the student to solve to find out how many more, e.g., How many more paper clips are there than erasers?
5. Demonstrate a counting up strategy.
6. Starting with the group that has fewer (e.g., erasers), and using the One Hundreds Chart, count up to the larger number (paper clips). Say: "There are 15 erasers and 22 paper clips." Point to the hundreds chart and count: "16, 17, 18, 19, 20, 21, 22. We counted 7 numbers. There are 7 more paper clips than erasers."
7. Continue procedure until the students are firm.

Note: Some students may find a counting back strategy to be easier.

Assign the worksheets: **Summer Sports 1** and **2**

Lesson 10

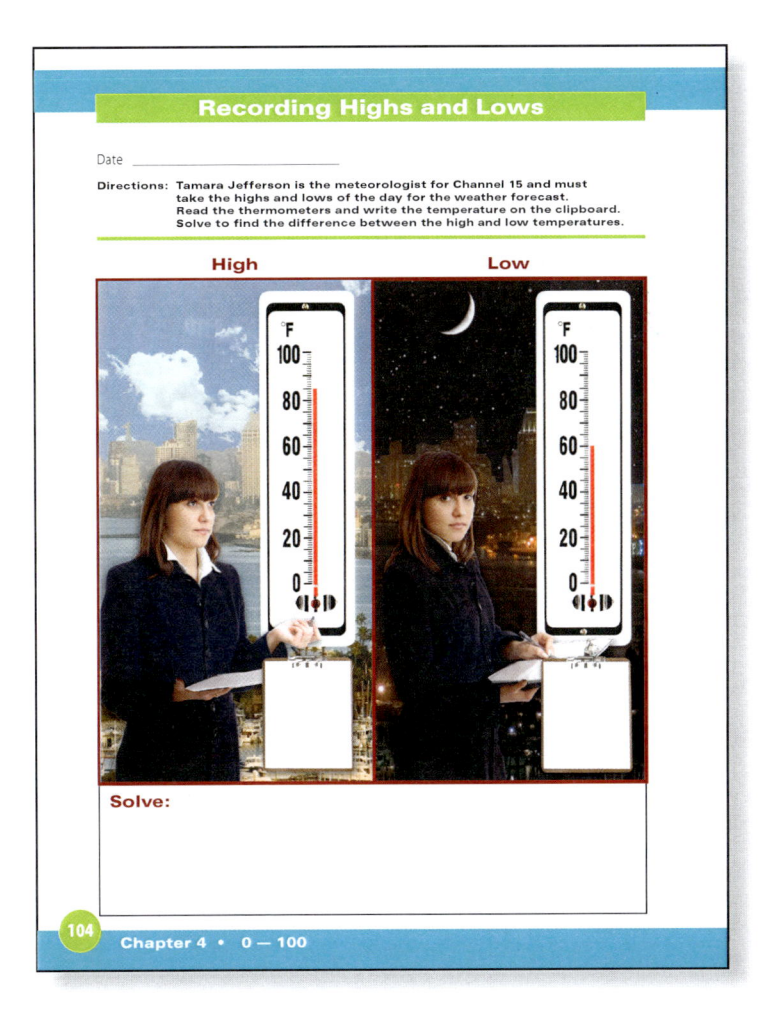

Objective

S. will find the difference between a high and low temperature.

Materials

- 2 teacher-made thermometers on the board, one labeled "high" and the other "low"
- dry board
- markers
- pencils
- student worksheet, p. 104: Recording Highs and Lows

Procedure

1. Point to the thermometer.

2. Review how to read a thermometer.

3. Review how to write "degrees" using the ° symbol.

4. Fill in each thermometer.

5. Ask a student to read the temperature.

6. Subtract the low temperature from the high temperature.

7. Remember to emphasize that the answer must be in degrees.

8. Lead the students through the task.

9. Continue with additional examples until the students are firm.

Assign the worksheet:
Recording Highs and Lows

Lesson 11

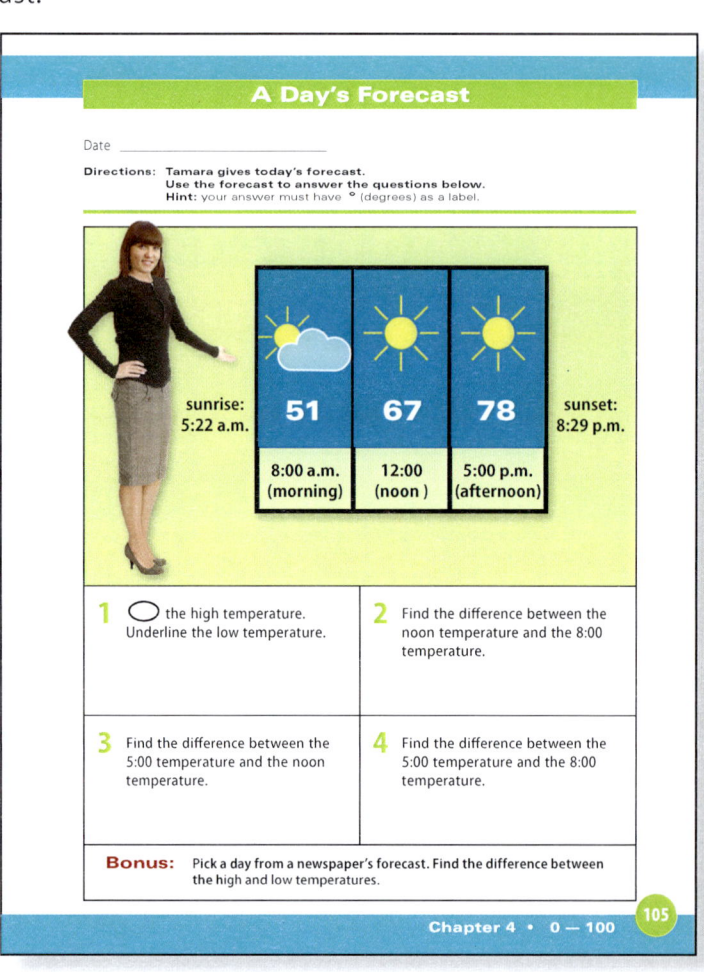

Objective

S. will find the difference between a high and low temperature.

Materials

- newspaper clippings of weather forecasts
- teacher-made daily temperature chart
- dry board
- markers
- pencils
- student worksheet, p. 105: A Day's Forecast

Procedure

1. Show the newspaper clippings.
2. Tell the students how to read the forecasts.
3. Point to the teacher-made forecast.
4. Compare it to the newspaper clippings.
5. Point to the teacher-made daily temperature chart.
6. Ask individual students to read the temperatures.
7. Demonstrate how to find the difference between the noon and morning temperatures.
8. Do the same procedure for all of the other temperatures.
9. Continue until students are firm.

Assign the worksheet:
A Day's Forecast

Lesson 12

Objective

S. will find the difference between the high and low temperatures.

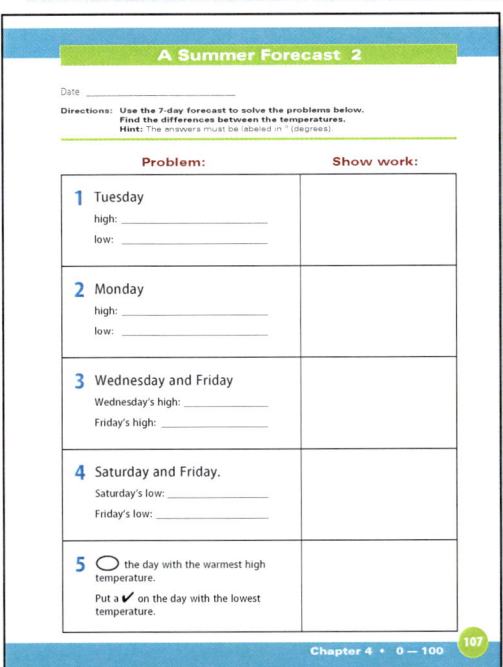

Materials

- teacher-made weeklong forecast
- dry board
- markers
- pencils
- student worksheets, pp. 106-107: A Summer Forecast 1 and 2

Procedure

1. Point to the weeklong forecast and call on individual students to read the temperatures for each day.

2. Find the highest/lowest temperature for the week.

3. Pick a day and find the difference between the high and low temperatures.

4. Continue until all of the days have been used.

5. Next, find the difference between the two high temperatures or two low temperatures.

6. Continue until students are firm.

Assign the worksheets:
A Summer Forecast 1 and **2**

Lesson 13

Objective

S. will solve subtraction word problems using money.

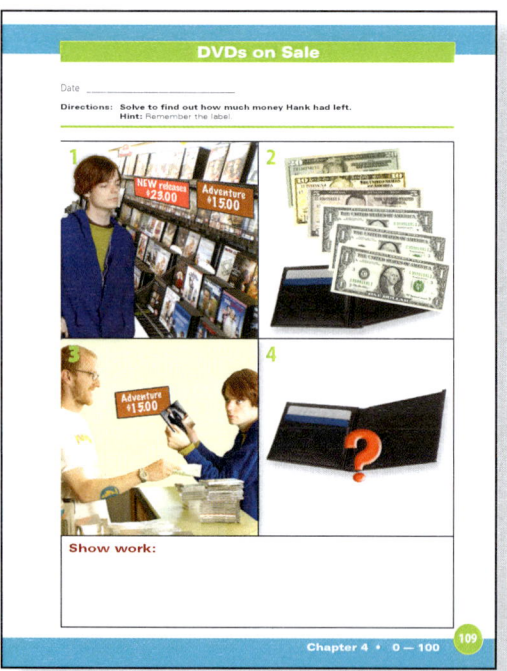

Materials

- set of items to buy with price tags attached (prices need to be in dollar amounts)
- sets of bills including twenties, tens, fives and ones.
- pencils
- student worksheets, pp. 108-109: A Shopping Trip and DVDs on Sale

Procedure

1. Lay the items to purchase in a row in front of the students.
2. Demonstrate how to purchase an item.
3. Count your money.
4. Select an item that cost less than the money in hand.
5. After buying the item, count the money left.
6. Lead the students through the steps.
7. Give student volunteers opportunities to purchase items and count the money left.
8. Continue until students are firm.

Assign the worksheets:
A Shopping Trip and **DVDs on Sale**

Lesson 14

Objective

S. will solve to find out a day's wage.

Materials

- analog or digital clock
- set of bills with various denominations
- pencils
- student worksheet, p. 110: All in a Day's Work

Procedure

1. Review how to tell time to the hour using the clock.

2. Review how to tell time to the hour using elapsed time, e.g., it is 9:00 and I worked 3 hours raking my yard. When did I stop working?

3. Next, create scenarios where someone gets paid per hour, e.g., Sam gets $8.00 an hour for delivering pizza. He started work at 5:00 and finished 2 hours later. Set the hands of the clock to 5:00. Move the hands one hour to 6:00. Sam earned $8.00 for one hour. Count out the money. Move the hands on the clock to the next hour. Sam earned another $8.00 for the second hour. Count out another 8 dollars. To find the total amount earned, count all of the money. Say: "Sam earned $16.00 for delivering pizzas for 2 hours."

4. Continue with additional scenarios until students are firm.

Assign the worksheet **All in a Day's Work**

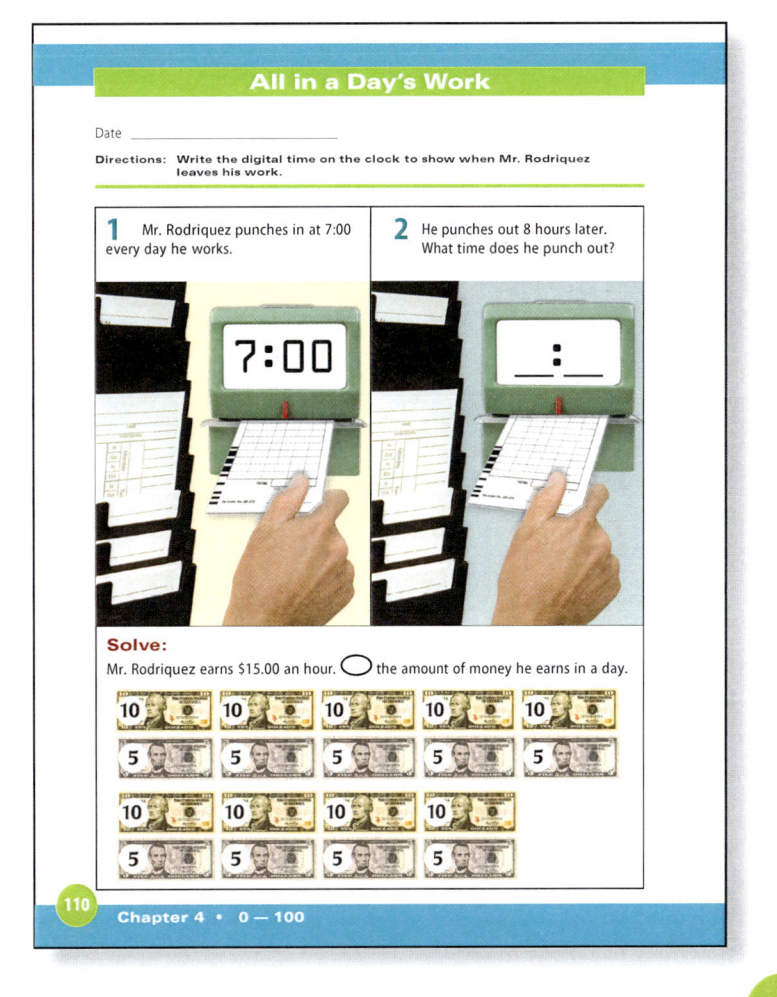

Chapter 4 • 0–100

Lesson 15

Objective

S. will draw hands on an analog clock that is thirty minutes later.

Materials

- analog clock(s)
- pencils
- student worksheet, p. 111: Mr. Rodriguez's Morning

Procedure

1. Review how to set the time to 30 minutes.

2. Create scenarios where students must find a time that is 30 minutes later, e.g., Tanya started her math homework at 7:00. Set the hands of the clock to 7:00. She finished her assignment 30 minutes later. At what time did she finish her homework? Count around the clock by 5 to 30 minutes. Set the clock at 7:30. Tanya finished her math homework at 7:30.

3. Create other scenarios calling on student volunteers to set the clock.

4. Now create scenarios where students must set the clock in a 30 minute sequence. For example, Philippe got up at 6:30, he ate breakfast 30 minute later. He was at the bus stop 30 minutes after he started to eat breakfast, etc.

5. Continue until students are firm.

Assign the worksheet:
Mr. Rodriguez's Morning

Note: For students to review counting by 5, use the One Hundreds Chart from Student Workbook, p. 90.

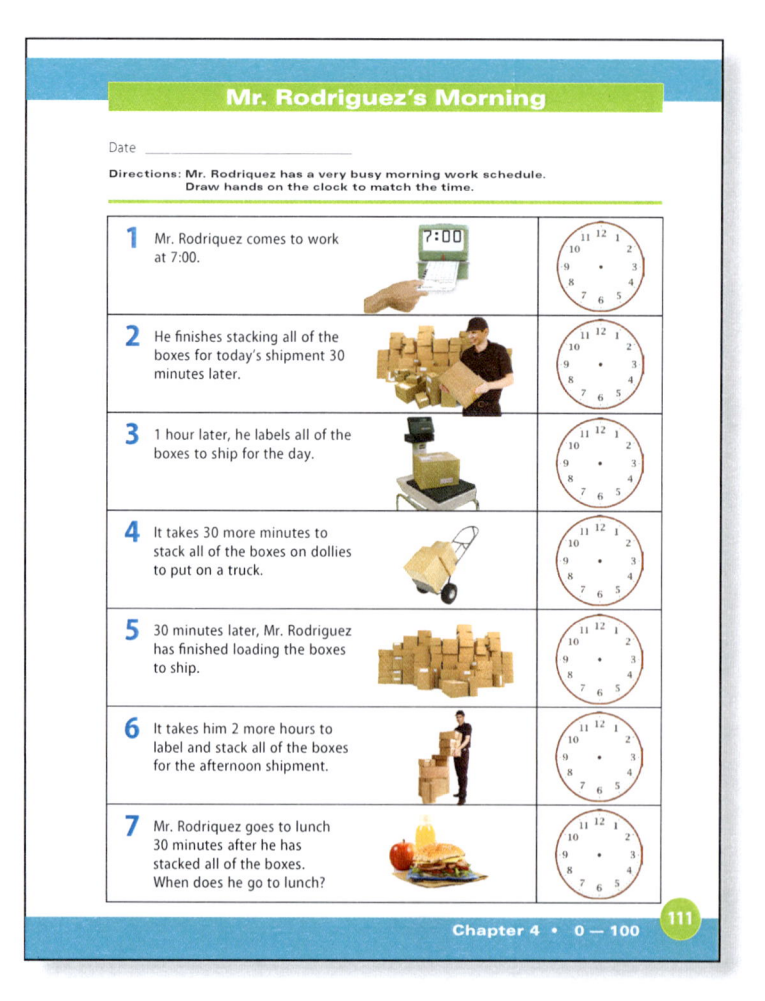

Lesson 16

Objective

S. will draw hands on an analog clock that is fifteen minute later.

Materials

- analog clock(s)
- pencils
- student worksheet, p. 112: 15 Minutes Later

Procedure

1. Review how to set the time to 15 minutes.

2. Create scenarios where students must find a time that is 15 minutes later. For example, Henry started to eat a snack at 8:00. He finished 15 minutes later. What time did he finish? Count around the clock by 5 to 15 minutes. Set the clock to 8:15. Henry finished eating at 8:15.

3. Continue with other scenarios, calling on student volunteers to set the clock.

4. Now create scenarios where students must set the clock in a 15 minute sequence, e.g., Martha started to text her friends at 9:00. She finished 15 minutes later. She received 5 messages 15 minutes after she finished texting, etc.

5. Continue until students are firm.

Assign the worksheet:
15 Minutes Later

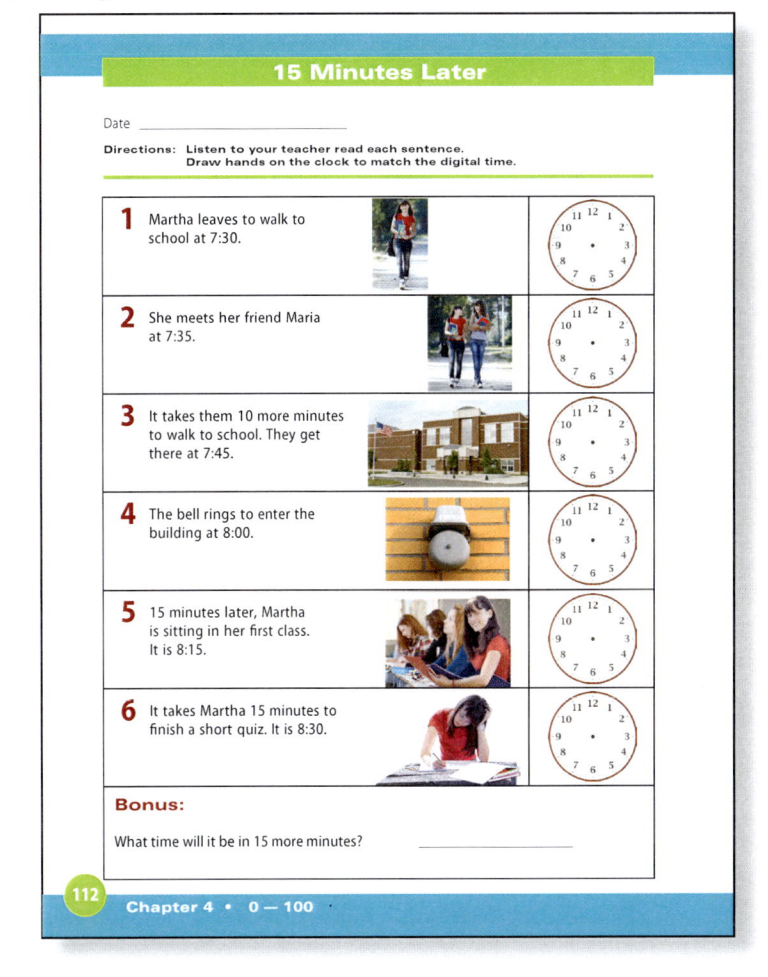

Chapter 4 • 0–100

Lesson 17

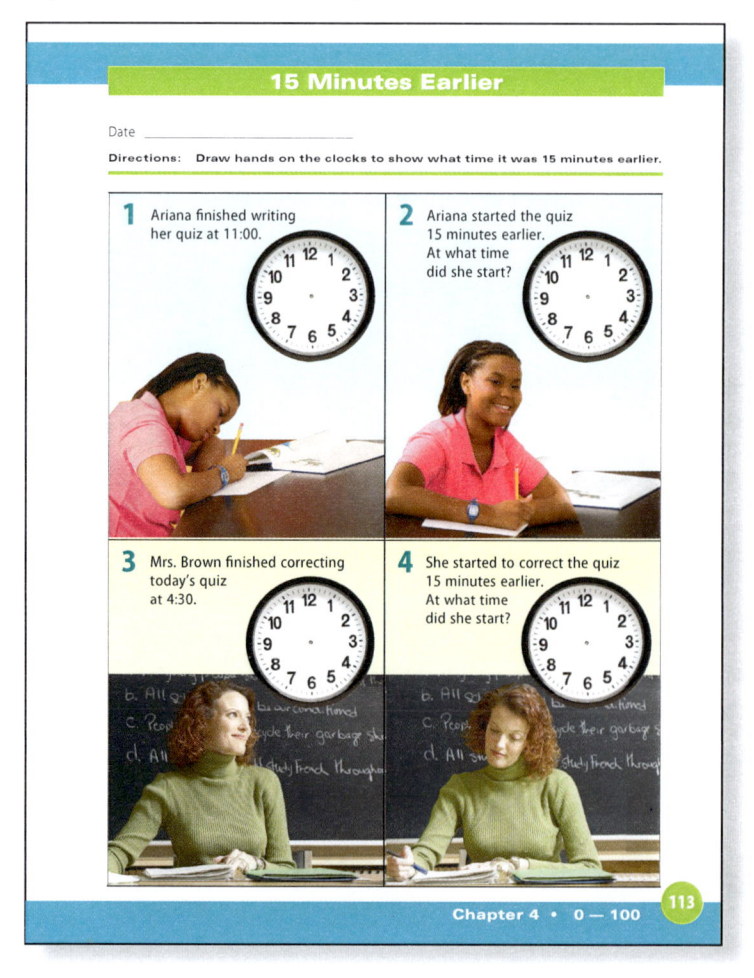

Objective

S. will draw hands on an analog clock that is fifteen minutes earlier.

Materials

- analog clock(s)
- pencils
- student worksheet, p. 113: 15 Minute Earlier

Procedure

1. Review how to set the clock to 15 minutes later.

2. Demonstrate how to set the clock to 15 minutes earlier.

3. Lead the students through the task.

4. Call on individual students to do the task.

5. Create a scenario using the concept of 15 minutes earlier, e.g., Shawano got home at 4:15. She left school 15 minutes earlier. At what time did she leave school? Count back 15 minutes. Set the clock to 4:00. Shawano left school at 4:00.

6. Continue until students are firm.

Assign the worksheet:
15 Minutes Earlier

Lesson 18

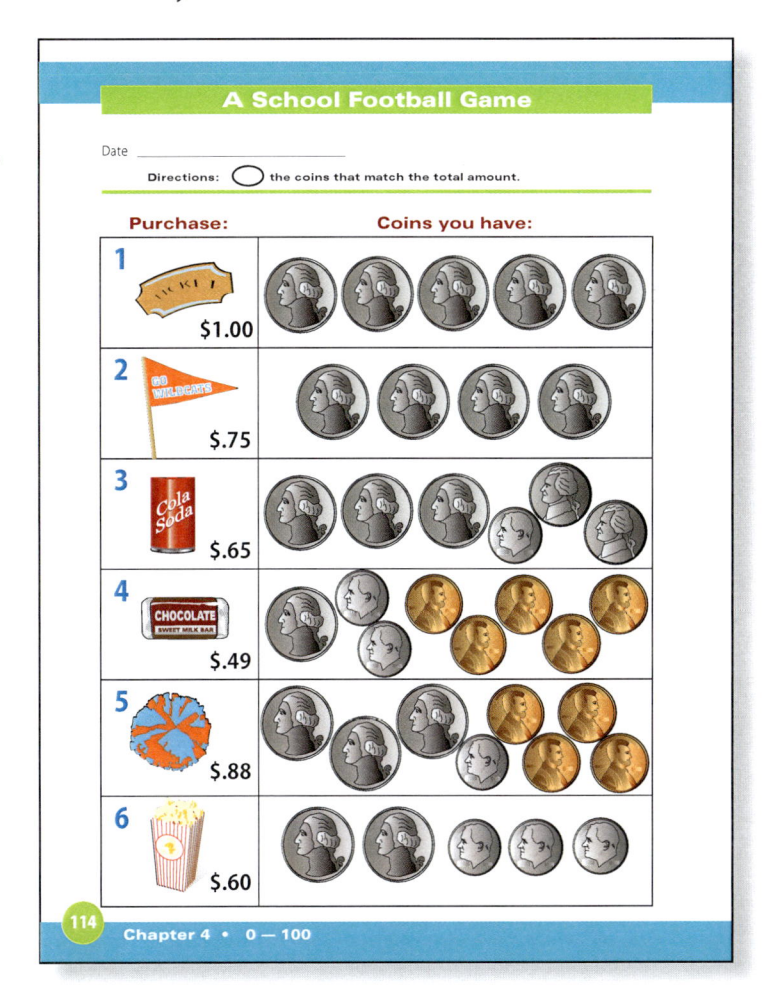

Objective

S. will match the total amount of coins to a purchase.

Materials

- small items to purchase with price tags to one dollar
- sets of pennies, nickels, dimes, and quarters to one dollar
- pencils
- student worksheet, p. 114: A School Football Game

Procedure

1. Place items in a row in front of the students.

2. Count out a set of money and use it to purchase an item.

3. The item cannot cost more than the money in the set.

4. Lead the students through the task.

5. Call on individual students to purchase an item.

6. Continue until students are firm.

Assign the worksheet:
A School Football Game

Lesson 19

Objective

S. will count back the change from one dollar.

Materials

- small items to purchase with price tags to one dollar
- sets of one dollar bills or four quarters
- cashier's drawer to count change
- pencils
- student worksheet, p. 115: Grocery Store

Procedure

1. Place items in a row to purchase.

2. Pick up a dollar bill and make a purchase.

3. Count change back, counting from the purchase price to one dollar.

4. Check your change to be sure that it is correct.

5. Assign student pairs, one to be the customer and the other the cashier.

6. The customer makes a purchase and hands the cashier one dollar in bills or coins.

7. The cashier counts back the change.

8. The customer checks to be sure that he receives the correct change back.

9. Continue until students are firm.

Assign the worksheet: **Grocery Store**

Lesson 20

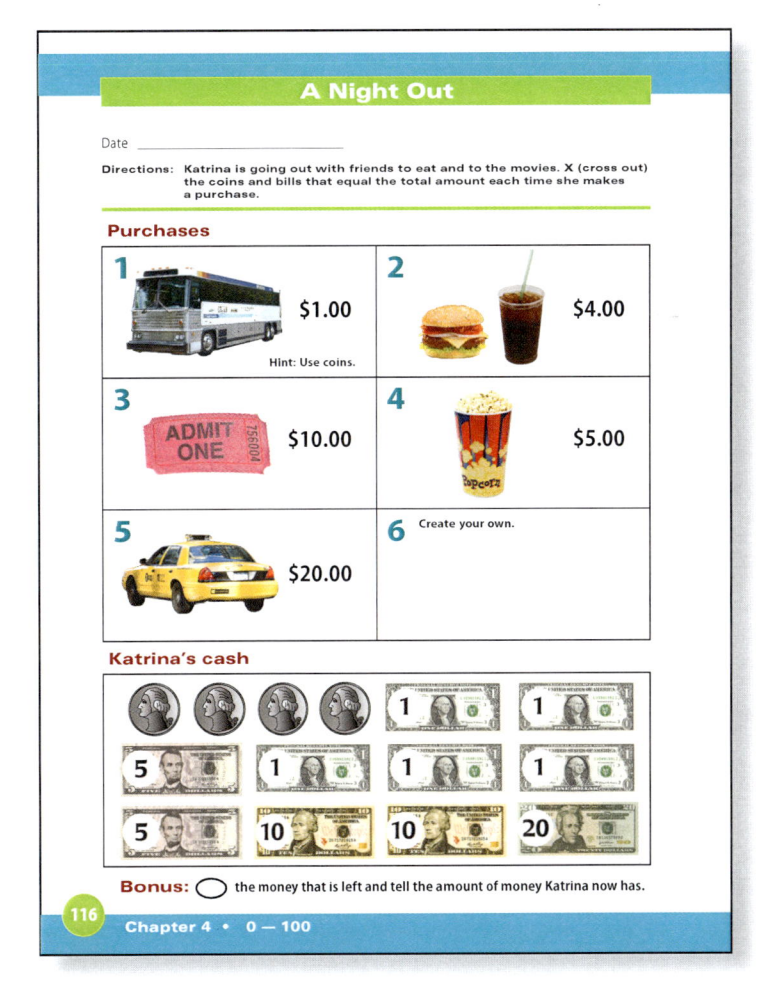

Objective

S. will match the total amount of coins/bills to the purchase price.

Materials

- items with price tags
- sets of bills/coins
- pencils
- student worksheet, p. 116: A Night Out

Procedure

1. Review counting coins/bills to make a purchase with the students.

2. Create a scenario where the students have a set amount of money to spend on multiple items, e.g., a trip to a mall, a movie, etc.

3. Demonstrate how to buy one item at a time until all of the items have been purchased.

4. Count the money left over.

5. Caution students that they can only purchase items as long as they have money to pay for them.

6. Continue with additional scenarios and until students are firm.

Assign the worksheet:
A Night Out

Lesson 21

Objective

S. will be able to count back change from a purchase price.

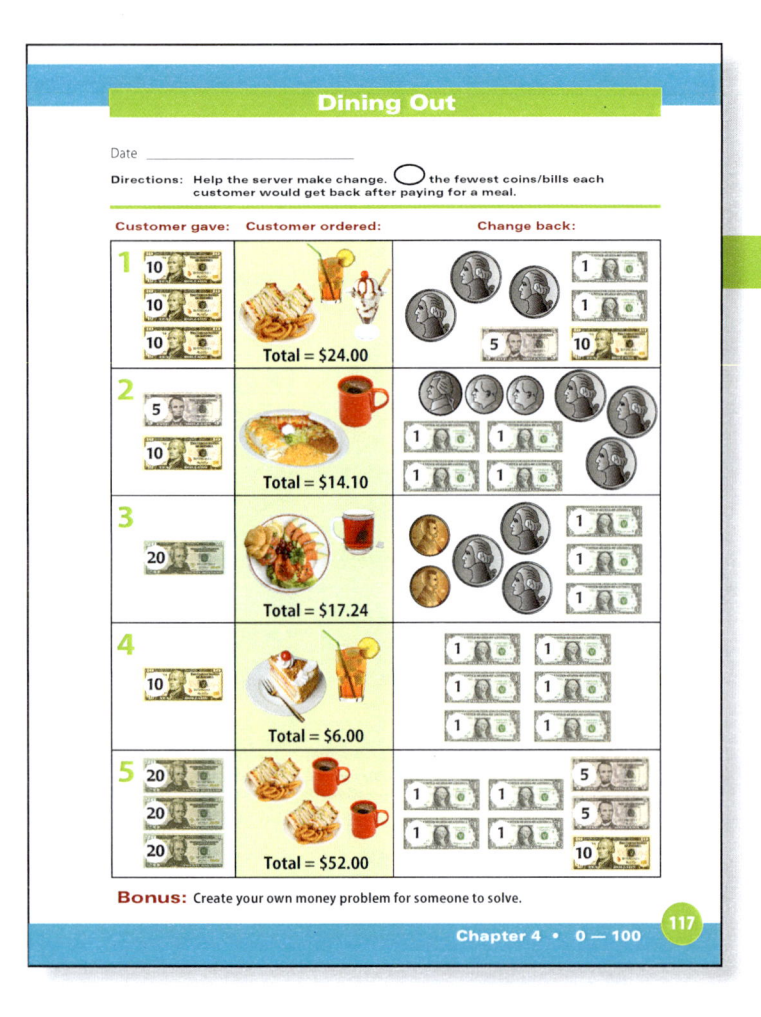

Materials

- teacher-made menu
- dry board
- markers
- sets of bills
- cashier's drawer with coins and bills
- pencils
- student worksheet, p. 117: Dining Out

Procedure

1. Demonstrate how to make more than one purchase from a menu.
2. Count the money in hand.
3. Select two or more items from the teacher-made menu.
4. Total the amount.
5. The amount cannot be more than the money.
6. Give the cashier the money.
7. Count change back starting from the total purchase to the amount given the clerk.
8. Check to be sure that the change is correct.
9. Assign student pairs, one to be the customer and the other the cashier.
10. The customer orders items from the menu and adds the total.
11. The customer gives the cashier the money to pay for the items.
12. The cashier counts back the change.
13. The customer checks the amount to be sure is it accurate.
14. Students switch roles.
15. Continue until the students are firm.

Assign the worksheet: **Dining Out**

Lesson 22

Objective

S. will interpret information on a line graph.

Materials

- a teacher-made line graph
- dry board
- markers
- pencils
- student worksheets, pp. 118-119: A Line Graph 1 and 2

Procedure

1. Point to the line graph on the dry board.

2. Explain to students that this is a different kind of graph. It is called a line graph.

3. Demonstrate how to read the graph.

4. Lead students through the task.

5. Ask questions based upon the graph.

6. Continue until the students are firm in the use of the graph.

Assign the worksheets: **A Line Graph 1** and **2**

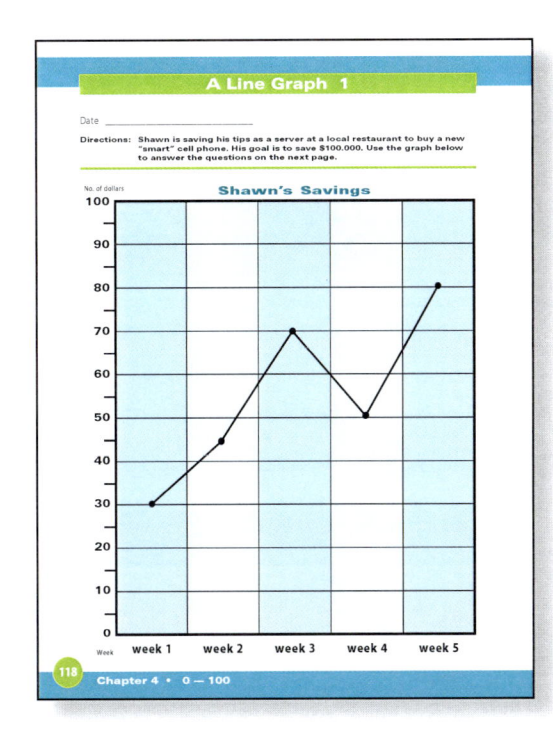

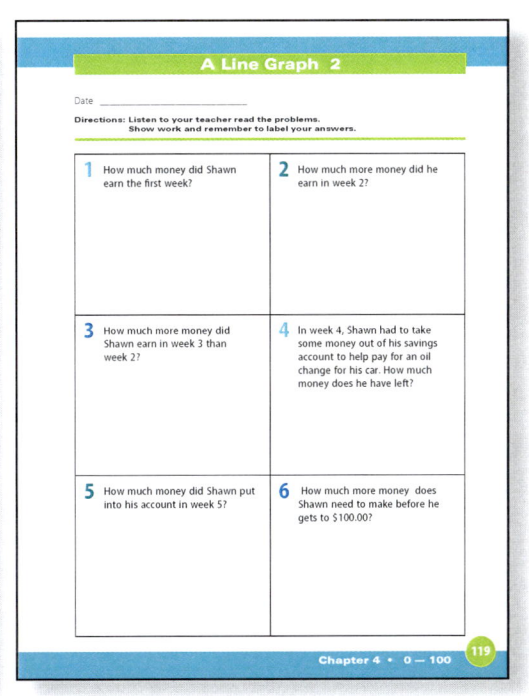

Lesson 23

Objective

S. will plot and interpret information on a bar graph.

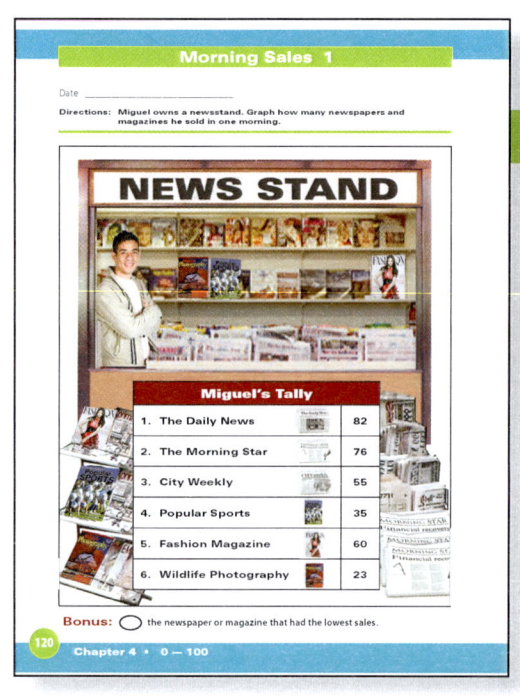

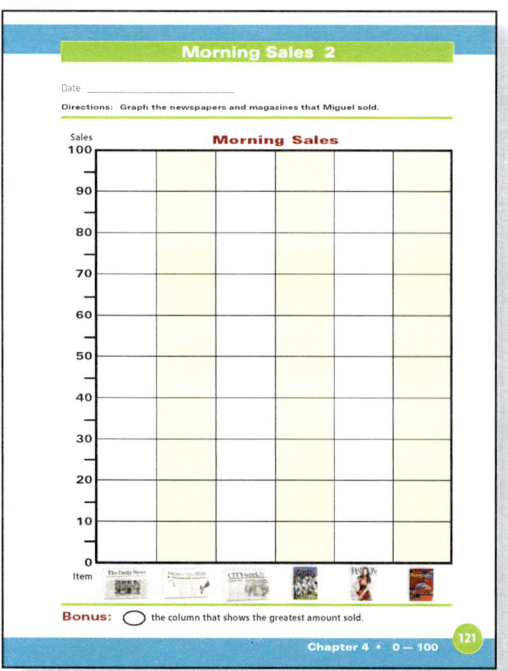

Materials

- teacher-made sales chart and bar graph
- dry board
- markers
- pencils
- student worksheets, pp. 120-121: Morning Sales 1 and 2

Procedure

1. Tell students that there are some jobs where people need to keep track of how many items they sell.

2. Brainstorm jobs where people would need to do this, e.g., store managers, salesmen, etc.

3. On a sales chart, list 5 items that are sold in a bookstore.

4. Beside the item, write a number that represents the amount sold that day.

5. Review the parts of the graph with the students.

6. Use the information from the sales chart to plot on the graph.

7. Demonstrate how to complete the first column.

8. Call on student volunteers to complete the next columns.

9. Ask questions based upon the information displayed on the graph.

10. Continue until students are firm.

Assign the worksheets:
Morning Sales 1 and 2

Explore Math Teacher's Manual

Lesson 24

Objective

S. will plot and interpret information from a bar graph.

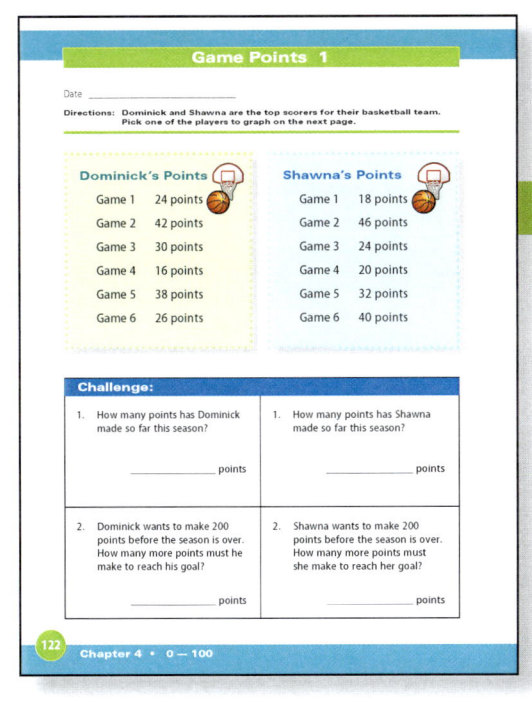

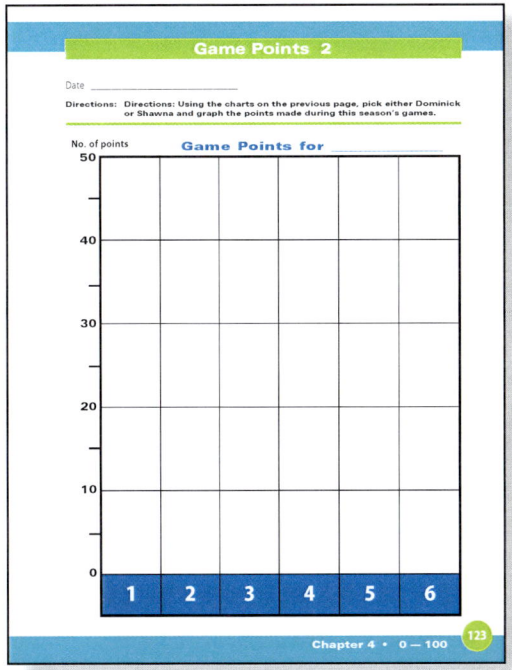

Materials

- teacher-made sheet showing how many points a player has made during a series of games
- teacher-made bar graph
- markers
- pencils
- student worksheets, pp. 122-123: Game Points 1 and 2

Procedure

1. Make a list of game points and games played. (See p. 121 in student workbook for an example.)

2. Ask questions about the game points, e.g., "In which game did the player score the most?" "How many points scored so far?"

3. Plot the total points for the first game.

4. Ask student volunteers to plot the next games until all of the games have been plotted.

5. Ask questions based upon the information on the graph.

6. Continue until students are firm.

Assign the worksheets:
Game Points 1 and **2**

Lesson 25

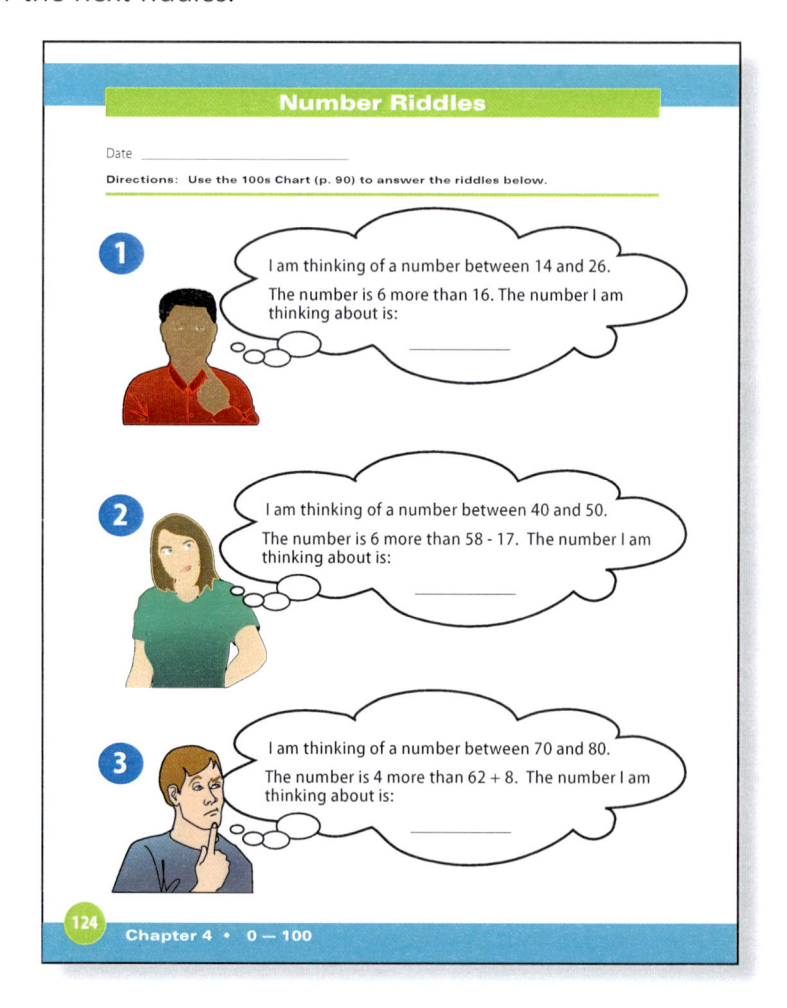

Objective

S. will use the one hundreds chart to answer a number riddle.

Materials

- One Hundreds Chart from Student Workbook, p. 90
- teacher-created number riddles
- pencils
- student worksheet, p. 124: Number Riddles

Procedure

1. Say a number riddle, e.g., "I am thinking of a number that is between 1 and 15. The number is 3 less than 12. What number am I?"

2. Demonstrate how to find the answer using the One Hundreds Chart.

3. Call on individual students for the next riddles.

4. Continue until students are firm.

Assign the worksheet:
Number Riddles

Number Riddles

Date _____

Directions: Use the 100s Chart (p. 90) to answer the riddles below.

1 I am thinking of a number between 14 and 26. The number is 6 more than 16. The number I am thinking about is: _____

2 I am thinking of a number between 40 and 50. The number is 6 more than 58 - 17. The number I am thinking about is: _____

3 I am thinking of a number between 70 and 80. The number is 4 more than 62 + 8. The number I am thinking about is: _____

Chapter 4 • 0 — 100

Explore Math Teacher's Manual

Lesson 26

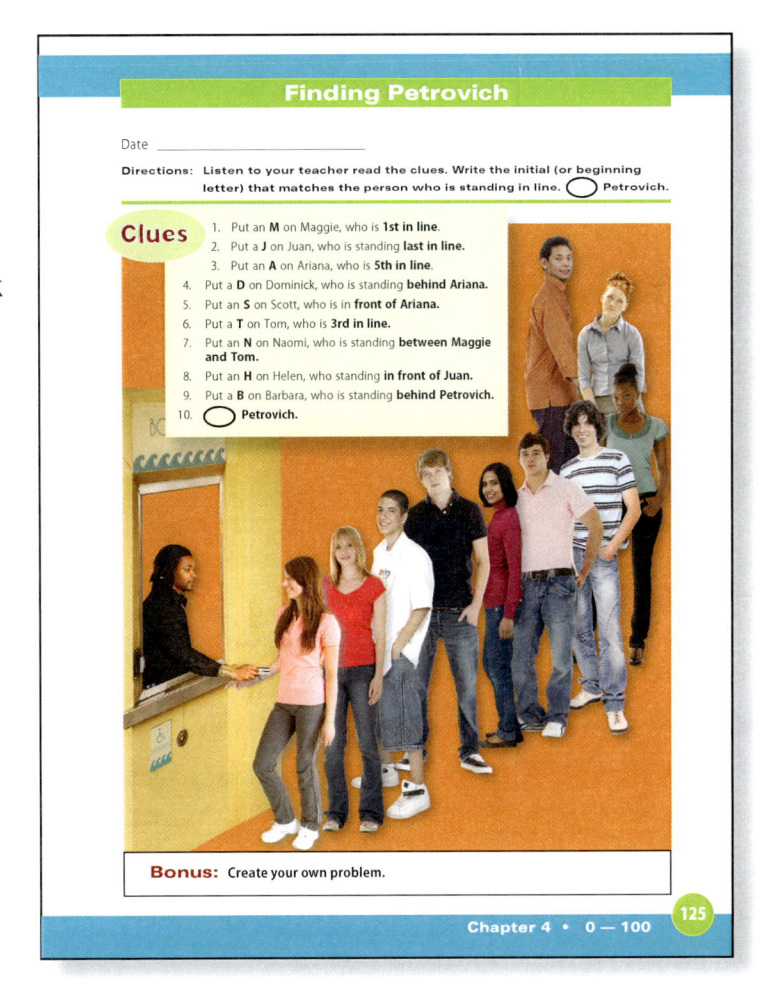

Objective

S. will solve a math riddle using clues.

Materials

- small manipulatives
- teacher-made math riddles using math clues and ordinal numbers
- dry board
- markers
- pencils
- student worksheet, p. 125: Finding Petrovich

Procedure

1. Lay the manipulatives out in a row in front of the students.

2. Read or say a math riddle.

3. Demonstrate how to solve the riddle by pointing to the manipulatives.

4. Lead the student through the task again.

5. Create another riddle and ask student volunteers to solve the riddle.

6. Continue until the students are firm.

Assign the worksheet:
Finding Petrovich

Chapter 5
0–1000

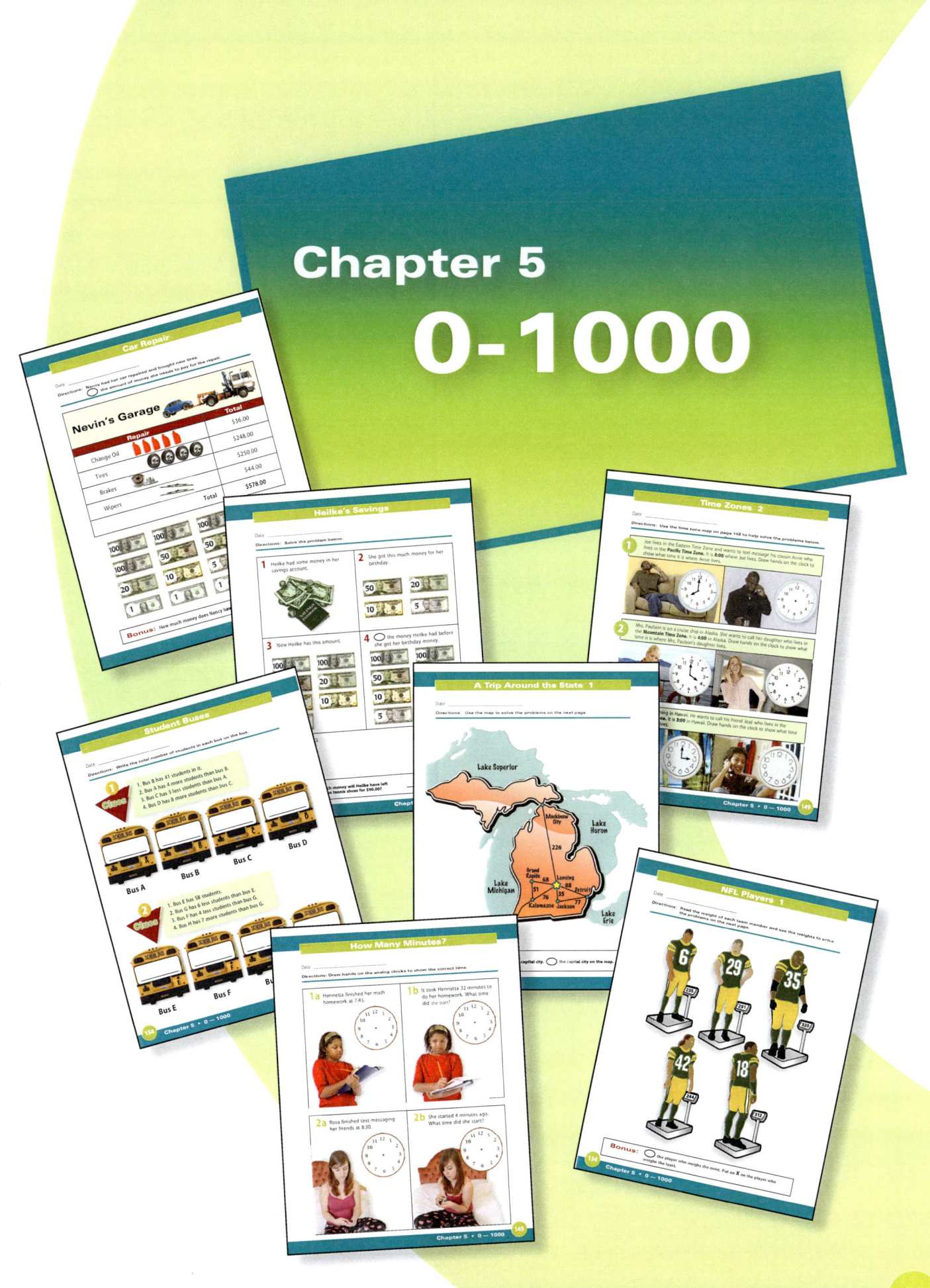

Lesson 1

Objective

S. will match the number of bills to the total price of an item.

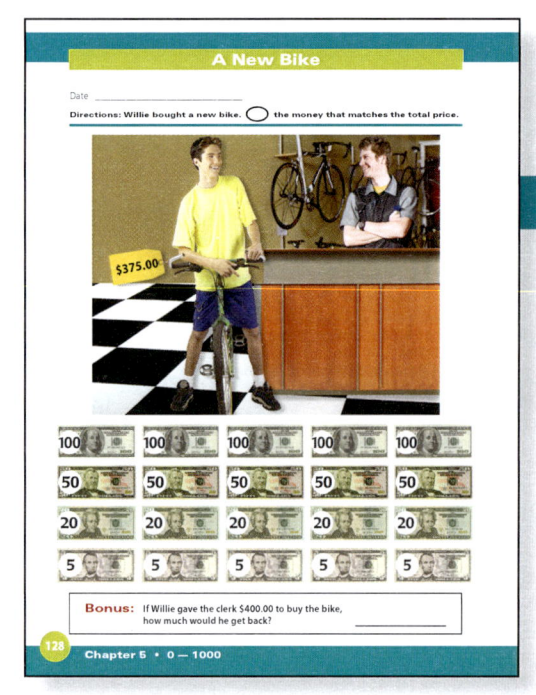

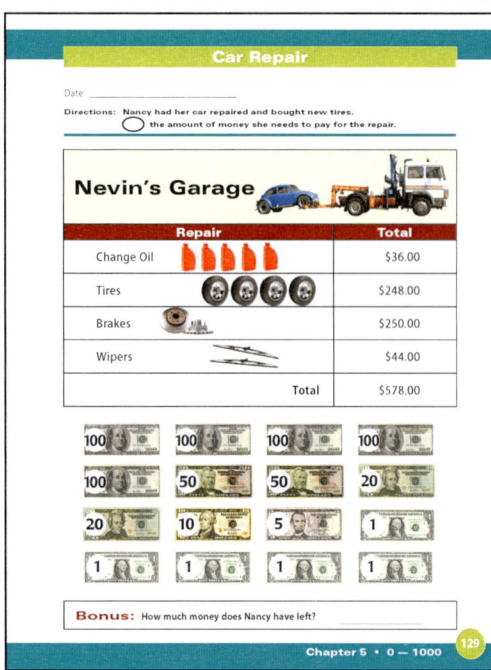

Materials

- sets of bills with denominations from one to one hundred
- items or pictures of items with price tags costing more than $100.00
- pencils
- student worksheets, pp. 128-129: A New Bike and Car Repair

Procedure

1. Students are to imagine going to a store that sells big ticketed items, e.g., an appliance or furniture store.

2. Lay the items in a row in front of the students.

3. Select a set of bills.

4. Count the bills and select an item that can be bought using the bills. Caution students not to purchase an item which costs more than the money in the set.

5. Count out the bills to match the purchase price.

6. Lead the students through the task.

7. Call on individual students to make a purchase.

8. Continue until students are firm.

Assign the worksheets:
A New Bike and **Car Repair**

Lesson 2

Objective

Using a map, S. will calculate the distance between two cities or towns.

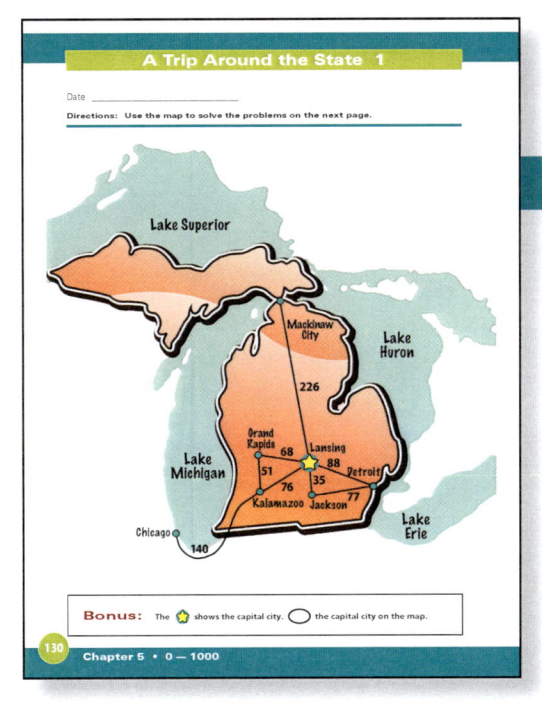

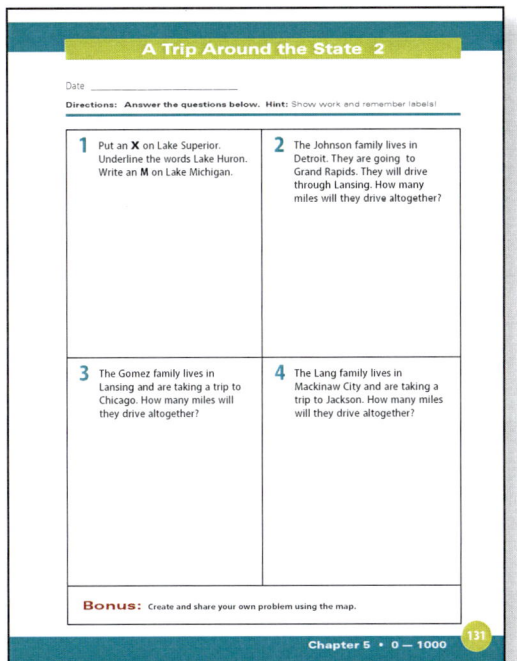

Materials

- state maps
- computer-generated maps
- teacher-made map on a dry board
- markers
- pencils
- student worksheets, pp. 130-131: A Trip Around the State 1 and 2

Procedure

1. Show state maps.

2. Explain special features.

3. Tell how to find the miles between two cities.

4. Point to the teacher-made map.

5. Describe the special features on the map.

6. Point out how to tell the distance in miles using the map.

7. Create a trip that a family would take going from one city to another.

8. Demonstrate how to use the miles on the map to add the total number of miles driven.

9. Continue with additional journeys until students are firm.

Assign the worksheet:
A Trip Around the State 1 and **2**

Lesson 3

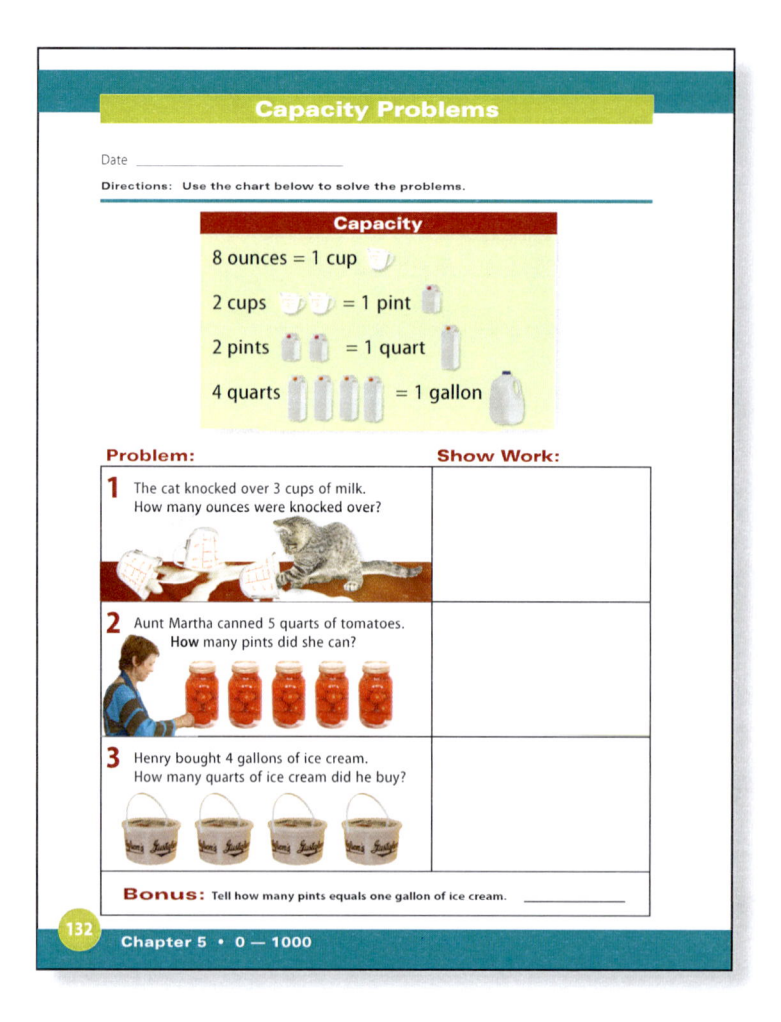

Objective

S. will convert standardized measurements, using capacity.

Materials

- measuring cups
- pint, quart, gallon containers
- sand or other substance to fill containers
- teacher-made chart showing conversions (see worksheet as an example)
- teacher-created capacity problems
- dry board
- markers
- pencils
- student worksheet, p. 132: Capacity Problems

Procedure

1. Give a definition of capacity, e.g., a measure of the amount a container can hold.

2. Show the students an example of how to measure capacity.

3. Point to and read the chart.

4. Give an example of each conversion.

5. Tell students that they are going to learn how to solve problems which require them to convert one measurement of capacity into another.

6. Give examples where students must practice converting different measurements, e.g., You have one pint. How many cups do you have?

7. Ask student volunteers to answer and check the answer using the pints and cups. Slowly increase the difficulty of each problem.

8. Continue until students are firm.

Assign the worksheet:
Capacity Problems

Explore Math Teacher's Manual

Lesson 4

Objective

S. will solve an addition problem using pounds.

Materials

- teacher-created weight problems
- dry board
- markers
- pencils
- student worksheet, p. 133: Weight Lifting

Procedure

1. Tell students that they are to imagine themselves in a gym.

2. Give various weight problems that individual students must "lift." For example, Jimmy lifted 15-pound weights, one in each hand. How many pounds did Jimmy lift? Sara lifted 60 pounds on each side of the weight lifting machine. How many pounds did Sara lift?

3. Demonstrate how to solve the problem.

4. Lead students through the task.

5. Continue until students are firm.

Assign the worksheet:
Weight Lifting

Lesson 5

Objective

S. will solve an addition problem using pounds.

Materials

- teacher-created weight problems
- dry board
- markers
- pencils
- student worksheets, pp. 134-135: NFL Players 1 and 2

Procedure

1. Ask students to give weights of NFL players.
2. Try to get at least five different weights.
3. Write the weights on the board.
4. Use the weights to create addition word problems which students must solve.
5. Students can check their answers with a calculator if necessary.
6. Continue until students are firm.

Assign the worksheets: **NFL Players 1** and **2**

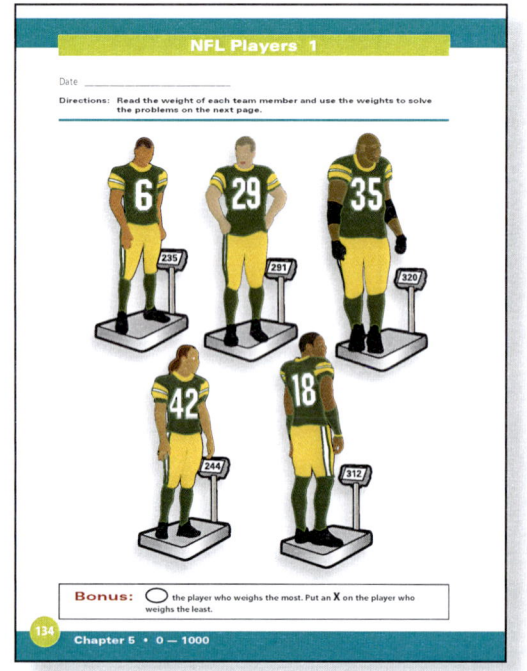

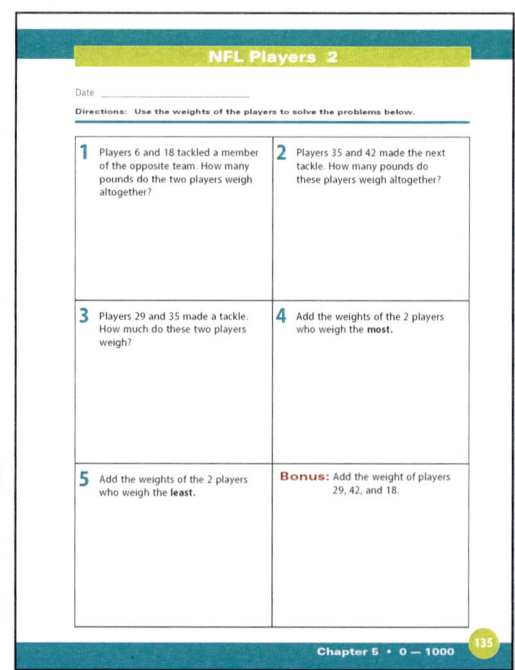

Lesson 6

Objective

S. will solve problems based upon a monthly budget.

Materials

- student/teacher-generated monthly budget
- dry board
- markers
- pencils
- student worksheets, pp. 136-137: A Monthly Budget 1 and 2

Procedure

1. With students, brainstorm what they think could be a good, part-time monthly wage.

2. Pick one.

3. Use the wage to create a monthly budget on the dry board.

4. Ask questions using the wage and monthly budget.

5. Continue until students are firm.

Assign the worksheets: **Monthly Budget 1** and **2**

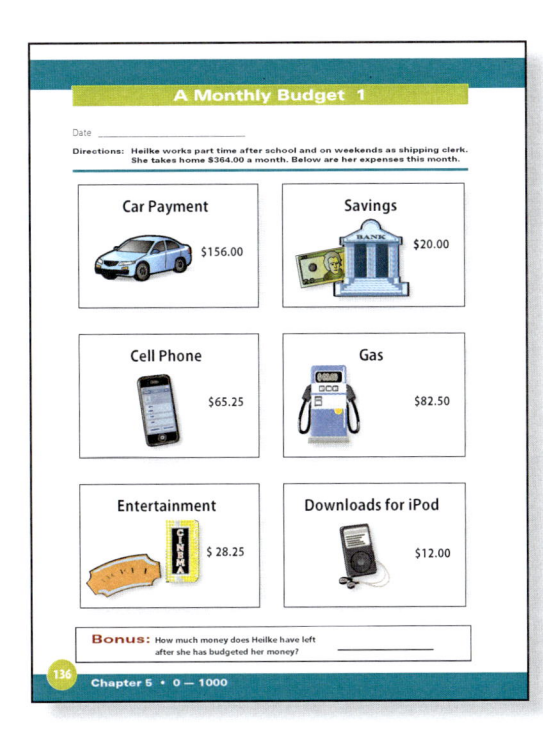

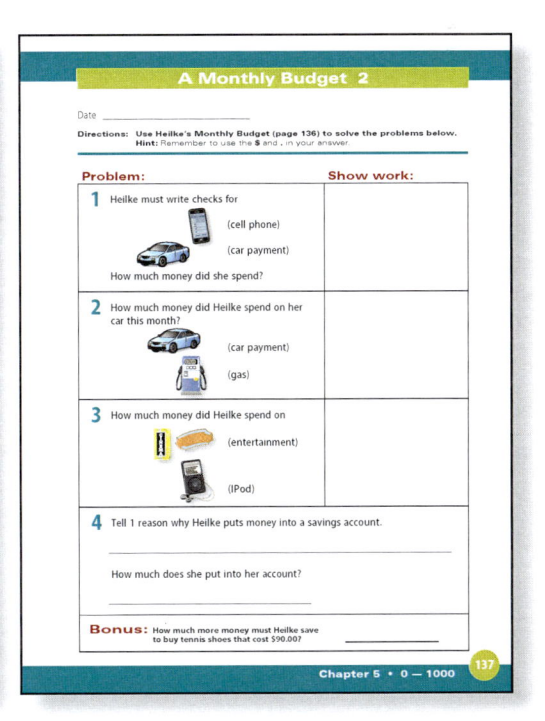

Lesson 7

Objective

S. will solve an addition word problem with a missing addend.

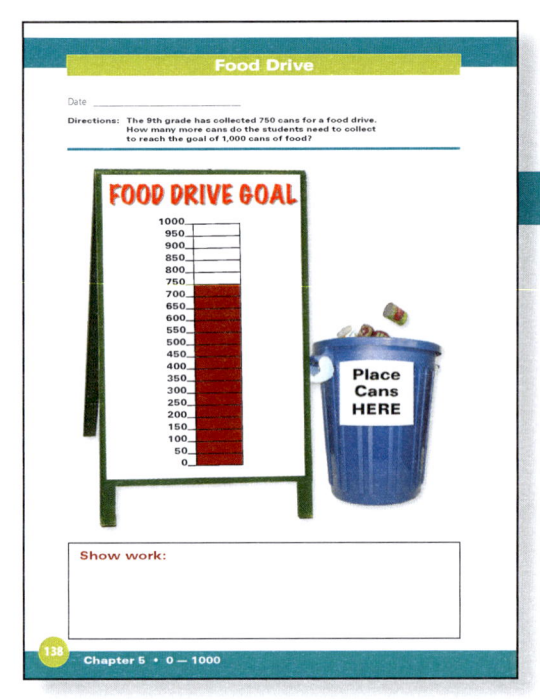

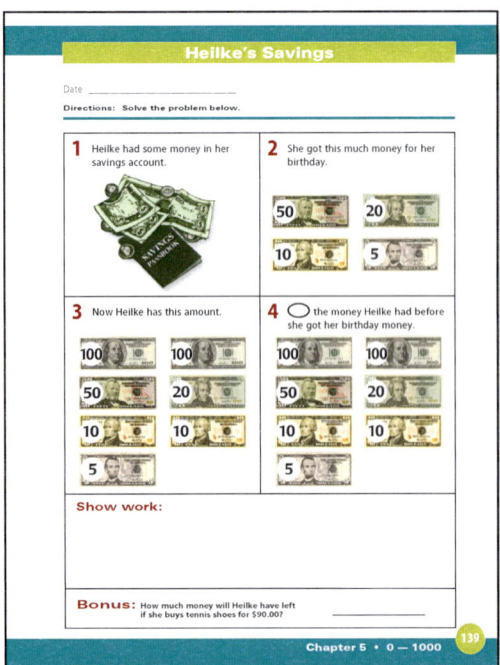

Materials

- teacher-made addition problems with one of the addends missing
- dry board
- markers
- pencils
- student worksheets, pp. 138-139: Food Drive and Heilke's Savings

Procedure

1. Write several addition problems on the board, each with a missing addend.

2. Point to the missing addend.

3. Create a scenario for the problem. Lead the students through the process of solving the problem.

4. Depending upon the students, use either a counting up or counting back strategy.

5. Continue until the students can solve the problem without prompts

Note: Students may need to use a calculator to solve these problems.

Assign the worksheets:
Food Drive and **Heilke's Savings**

Lesson 8

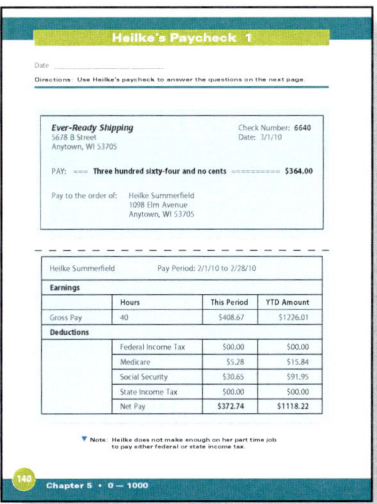

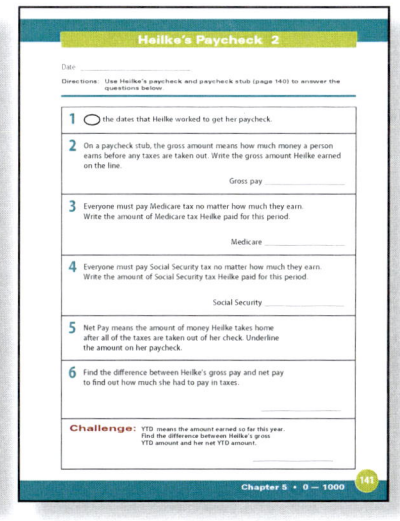

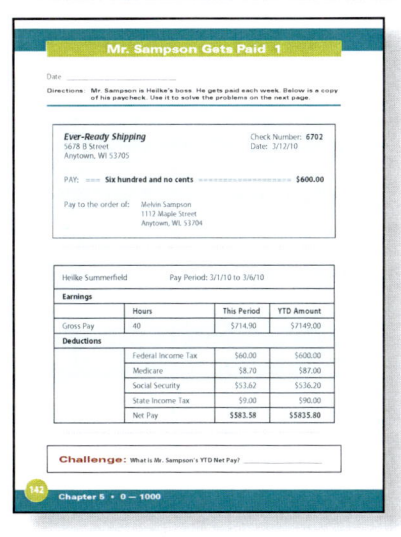

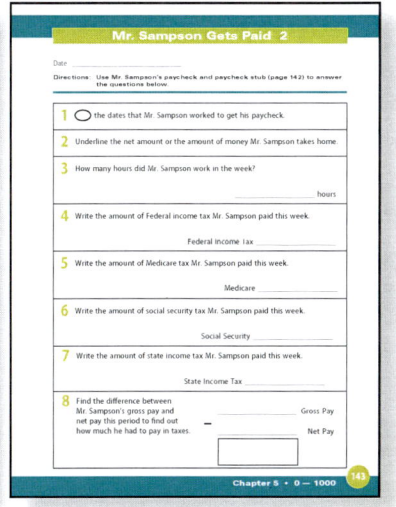

Objective

S. will solve problems based upon a paycheck and paycheck stub.

Materials

- teacher-made paycheck on board
- dry board
- markers
- pencils
- student worksheets, pp. 140-143: Heilke's Paycheck 1 and 2; Mr. Sampson Gets Paid 1 and 2

Procedure

1. Point to the different parts of the paycheck and stub, explaining what each part means.

2. Ask questions based upon the information shown on the board.

3. Create problems that the student must solve based upon the information shown.

4. Continue until the students are firm.

Assign the worksheets:
Heilke's Paycheck 1 and **2**
Mr. Sampson Gets Paid 1 and **2**

Lesson 9

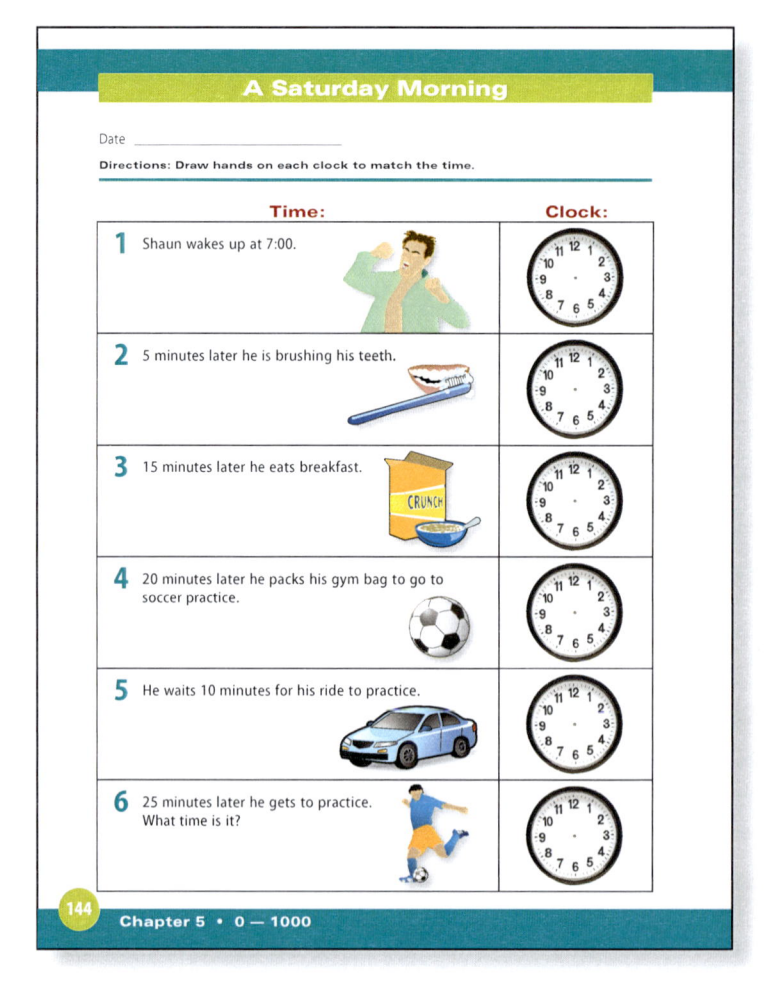

Objective

S. will draw hands on an analog clock that match the time.

Materials

- analog clock(s)
- digital times on board
- dry board
- markers
- pencils
- student worksheet, p. 144: A Saturday Morning

Procedure

1. Review how to read digital times to the hour, half-hour, quarter-hour, and 5 minutes.

2. Review how to set an analog clock to match the digital times.

3. Create scenarios where the students set an analog clock to a set time, then set the clock to a later time. For example, Timothy had art class at 9:00. He finished his art assignment 30 minutes later. What time did Timothy finish his project? Timothy finished cleaning his brushes 10 minutes after he finished his project. What time did he finish cleaning his brushes?

4. Continue until students are firm.

Assign the worksheet:
A Saturday Morning

A Saturday Morning

Date _____

Directions: Draw hands on each clock to match the time.

#	Time:	Clock:
1	Shaun wakes up at 7:00.	
2	5 minutes later he is brushing his teeth.	
3	15 minutes later he eats breakfast.	
4	20 minutes later he packs his gym bag to go to soccer practice.	
5	He waits 10 minutes for his ride to practice.	
6	25 minutes later he gets to practice. What time is it?	

Chapter 5 • 0 — 1000

Lesson 10

Objective

S. will draw hands on an analog clock to the correct time.

Materials

- analog clock(s)
- teacher-made list of digital times to the hour
- dry board
- markers
- pencils
- student worksheet, p. 145: How Many Minutes?

Procedure

1. Review how to tell time to the minute, by matching a digital time to a time on the analog clock.

2. Create scenarios where the student must set a clock minutes later than a digital time, e.g., point to 7:00 and say: "Melinda fed her cat at 7:00; she finished feeding it 2 minutes later. What time did Melinda stop feeding her cat?"

3. Demonstrate how to find minutes later. Start at 7:00 and count, "1, 2."

4. Melinda finished feeding her cat at 7:02.

5. Continue until the students are firm.

6. Follow the same procedure for minutes earlier.

Assign the worksheet:
How Many Minutes?

Lesson 11

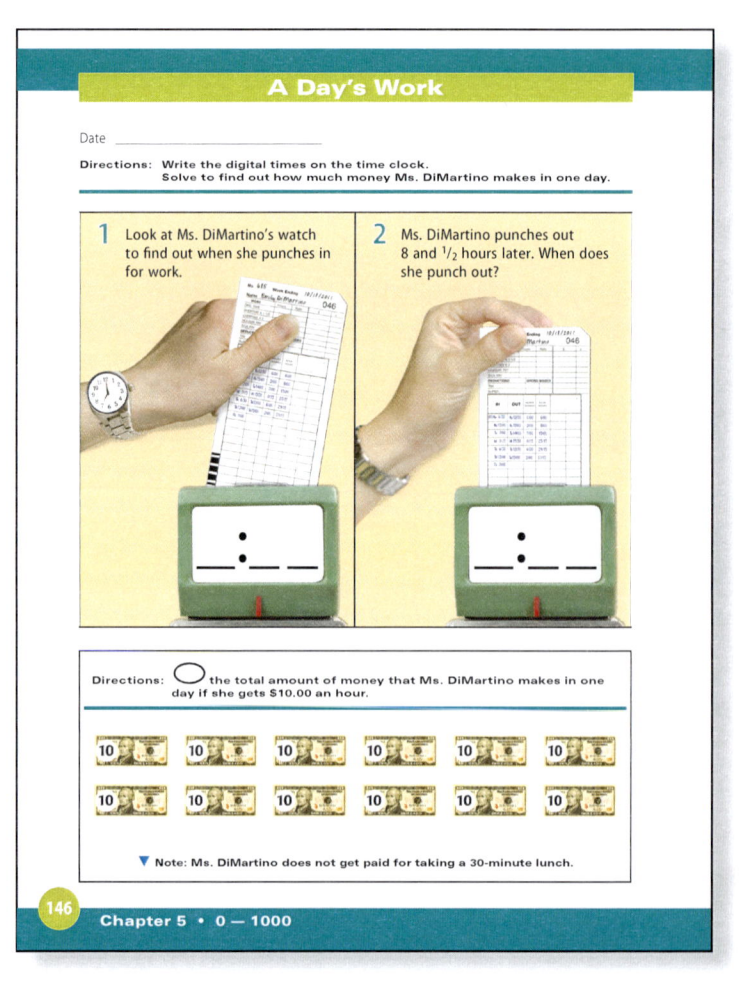

Objective

S. will solve to find out a day's wage.

Materials

- nalog clock
- set of bills with various denominations
- teacher-made list of digital times
- pencils
- student worksheet, p. 146: A Day's Work

Procedure

1. Review how to tell time by matching the digital time to the analog clock.

2. Point to a digital time and review how to find elapsed time using the analog clock. For example, point to 6:00 and say, "Simone is going to baby-sit. She started at 6:00 and baby-sat for five hours. When did she finish?" Count around the analog clock five hours. Simone finished at 11:00.

3. Next, create scenarios where someone gets paid per hour of work, e.g., if Simone gets $5.00 an hour to baby-sit, how much money did she make in five hours? Move the clock one hour from 6:00 to 7:00. Count out five dollars. Continue until it is 11:00. Count all the money. Simone made $25.00 for five hours of work.

4. Create other scenarios and continue until the students are firm.

Assign the worksheet:
A Day's Work

Lesson 12

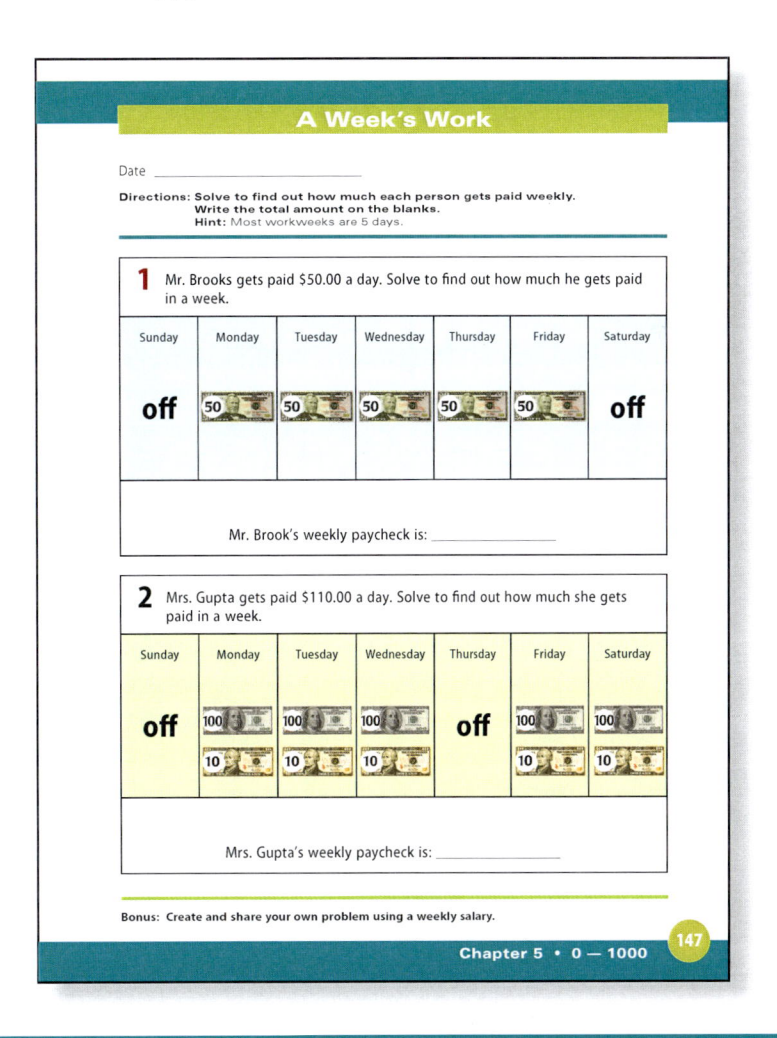

Objective

S. will solve to find out a week's wage.

Materials

- calendar(s)
- set of bills of various denominations
- pencils
- student worksheet, p. 147: A Week's Work

Procedure

1. Review how to read the days of the work.

2. Review the difference between a workweek and the calendar week with students.

3. Caution students that not all people who work get off both days of the weekend.

4. The workweek can include a weekend day(s).

5. Create scenarios where various people earn so much a day and the students must find out how much each person earns in a workweek, e.g., Mr. Severson earns $40.00 a day; how much does he earn in a week? Lay $40.00 on each day that Mr. Severson works and count out the total amount. Mr. Severson earns $200.00 a week.

6. Lead students through the task.

7. Continues until students are firm.

Assign the worksheet:
A Week's Work

Lesson 13

Objective

S. will solve to find the difference in U.S. time zones.

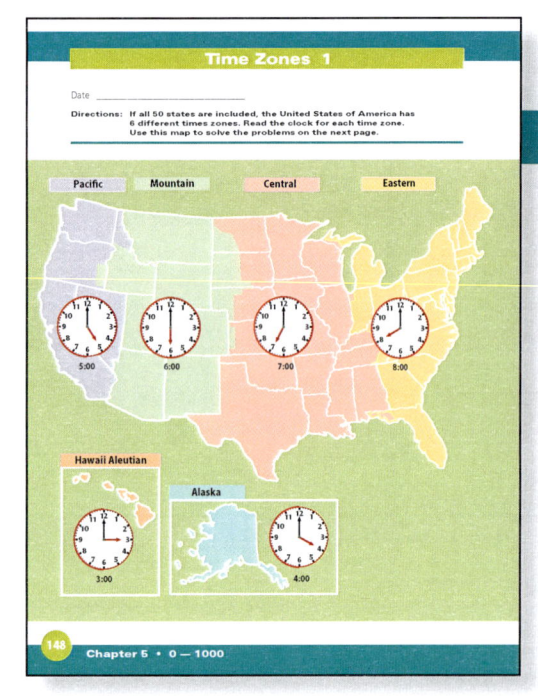

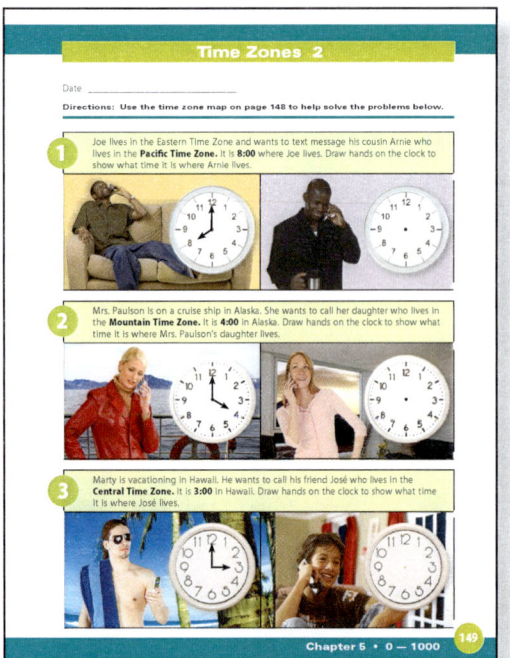

Materials

- map of all fifty states with time zones found in the U.S.
- analog clock for each time zone
- pencils
- student worksheets, pp. 148-149: Time Zones 1 and 2

Procedure

1. Point to the map and explain that if all fifty states are included, there are six different time zones.

2. Point to and talk about each time zone.

3. Talk about how you subtract an hour going from the Eastern Time zone to the Hawaii Aleutian Time Zone, and how you add an hour going in the opposite direction.

4. Create problems where students must use the time zones to find differences between two of the time zones, e.g., it is 4:00 in the Central Time zone. Natasha is going to call her friend Ravi, who lives the Pacific Time zone. What time is it in the Pacific Time Zone?

5. Point to and count back as the time zones are counted.

6. Set the analog clock in the Pacific Time zone to 2:00. It is 2:00 o'clock in Ravi's time zone.

7. Continue until the students are firm.

Assign the worksheets: **Time Zones 1** and **2**

Lesson 14

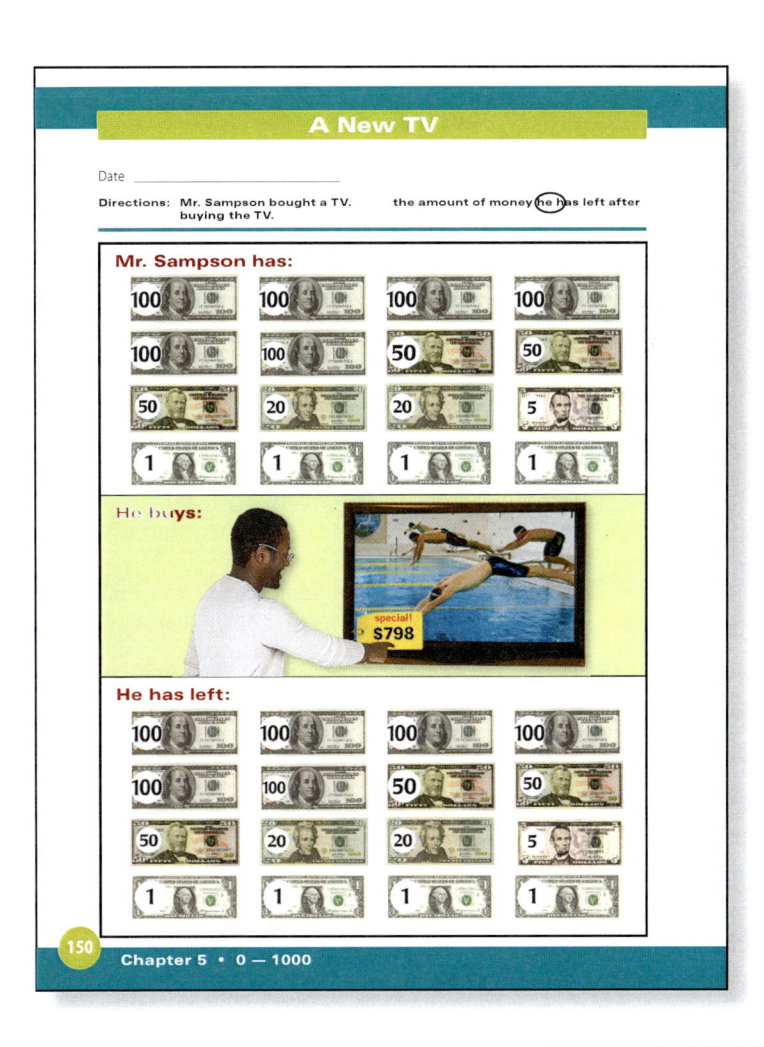

Objective

S. will match the total amount of bills to a purchase price.

Materials

- items with price tags
- sets of bills with various denominations
- pencils
- student worksheet, p. 150: A New TV

Procedure

1. Review with students how to count bills with different denominations to one thousand.

2. Demonstrate how to make a purchase.

3. Select a set of bills.

4. Select an item to buy.

5. Caution students that they must have more than enough money to pay for the item.

6. Count out the bills.

7. If there is more than enough money to pay for the item, count out the bills until the amount matches the purchase price.

8. Repeat.

9. Continue until students are firm.

Assign the worksheet:
A New TV

Lesson 15

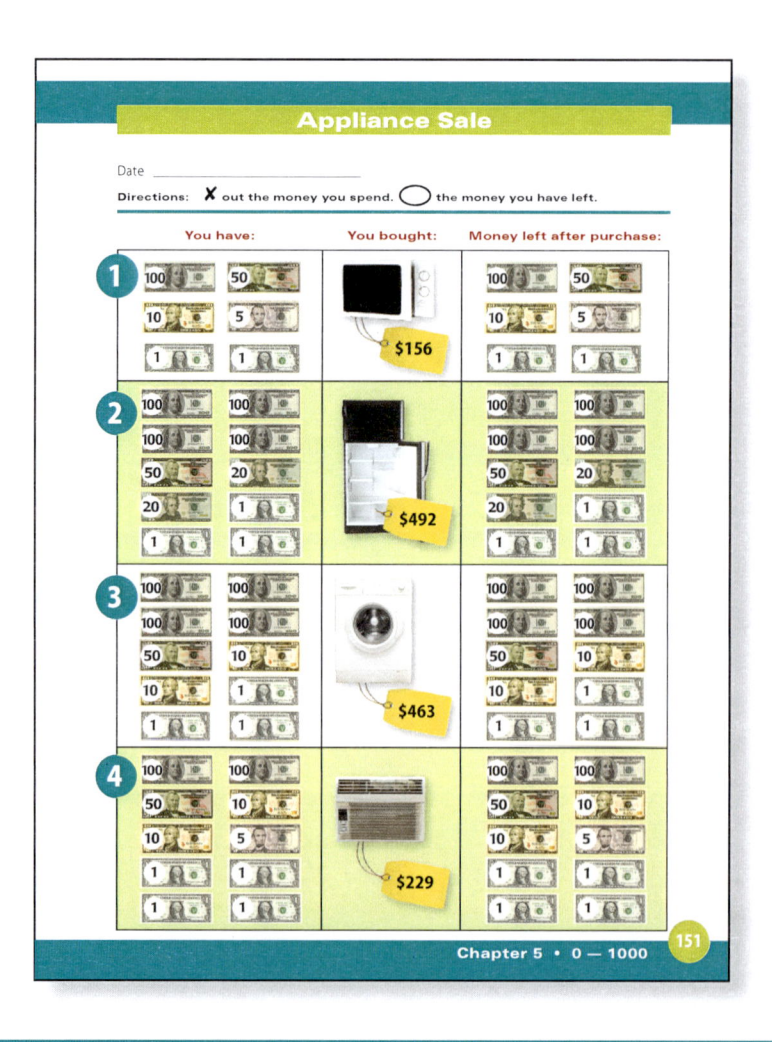

Objective

S. will count back change after making a purchase.

Materials

- items with price tags
- sets of bills with various denominations
- pencils
- student worksheet, p. 151: Appliance Sale

Procedure

1. Select a set of bills and count the money.

2. Caution students that they must have more than enough money to pay for the item. Select an item to purchase.

3. Give the cashier the money.

4. The cashier counts the change back, starting with the purchase price.

5. Check to be sure the change has been counted correctly.

6. Repeat procedure.

7. Assign student pairs, one to be a customer and the other the clerk. The customer buys an item and the clerk counts back the change.

8. Students switch roles.

9. Continue until the students are firm.

Assign the worksheet:
Appliance Sale

Lesson 16

Objective

S. will plot and interpret information on a bar graph.

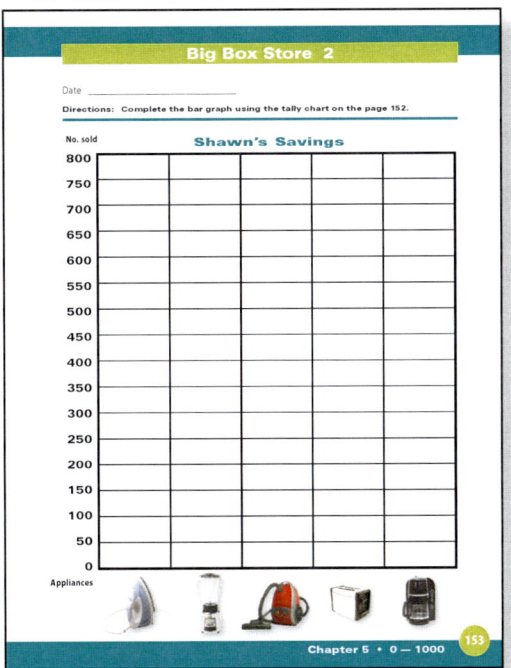

Materials

- teacher-made inventory chart (see p. 140 as an example)
- teacher-made bar graph
- pencils
- student worksheets, pp. 152-153: Big Box Store 1 and 2

Procedure

1. Review the parts of the graph.
2. Tell students that there are some jobs where people must keep track of inventory.
3. Explain what that means.
4. Brainstorm jobs where people would need to do this, e.g., store managers, restaurant managers, warehouse shippers, etc.
5. Create a scenario based upon one of the jobs.
6. Create an inventory chart.
7. Demonstrate how to use the information to create a bar graph.
8. Fill in the first column.
9. Ask student volunteers to fill in the other columns.
10. Ask questions based upon the information shown on the graph.
11. Continue until students are firm.

Assign the worksheet: **Big Box Store 1** and **2**

Lesson 17

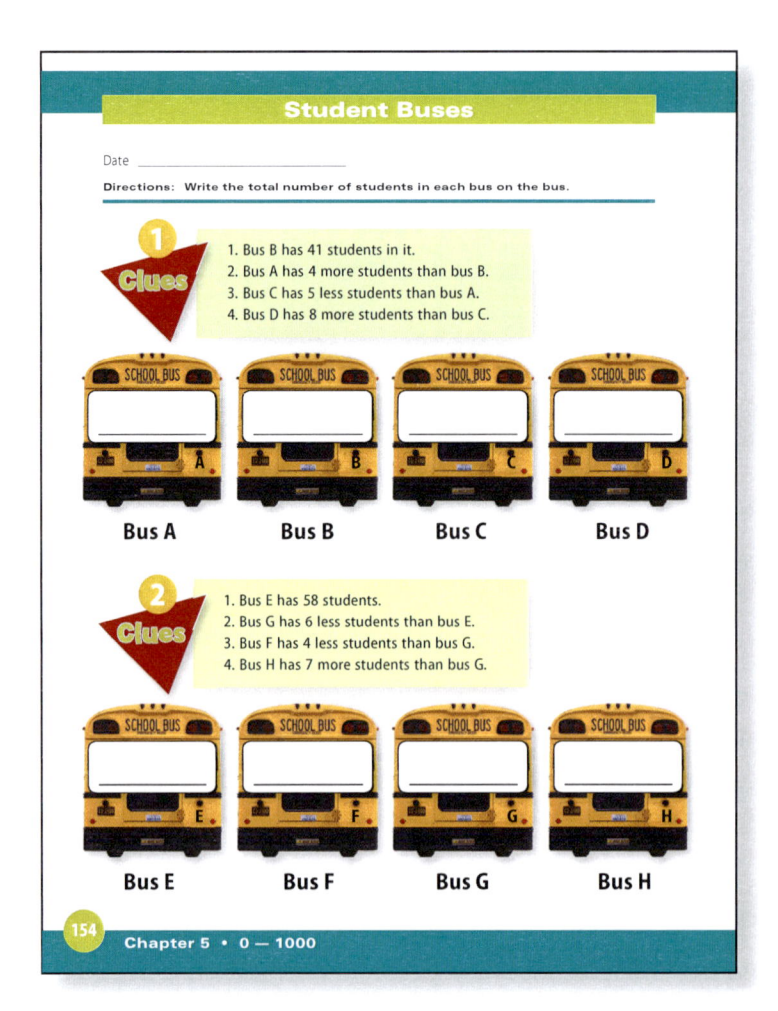

Objective

S. will solve a number riddle using the One Hundreds Chart.

Materials

- One Hundreds Chart, from Student Workbook, p. 90
- teacher-made riddles
- pencils
- student worksheet, p. 154: Student Buses

Procedure

1. Say or write a number riddle.

2. Demonstrate how to find the answer using the One Hundreds Chart, e.g., bus A has 60 people in it and bus B has 4 fewer people. How many people does bus B have? Bus C has 3 more people than bus B. How many people does bus C have?

3. Continue until students are firm.

Assign the worksheet: **Student Buses**

Lesson 18

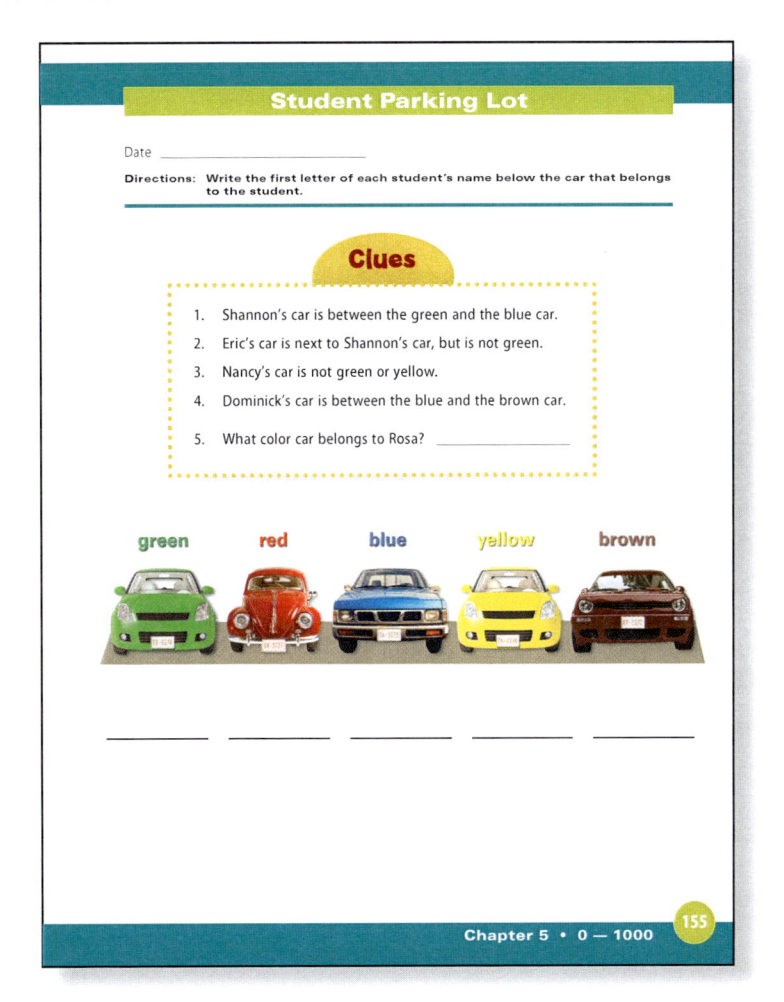

Objective

S. will solve a math riddle using clues.

Materials

- teacher-made riddles
- dry board
- markers
- pencils
- student worksheet, p. 155: Student Parking Lot

Procedure

1. Point to the drawing on the board, e.g., 5 different people. Say: "Sara is standing between the student who is wearing yellow and the one wearing black. Put an **S** on Sara. Ralph is next to Sara but is not wearing yellow. Write an **R** on Ralph, etc.

2. Continue until all 5 students have been identified.

3. Repeat using other riddles until students are firm.

Assign the worksheet:
Student Parking Lot

Student Parking Lot

Date _____

Directions: Write the first letter of each student's name below the car that belongs to the student.

Clues

1. Shannon's car is between the green and the blue car.
2. Eric's car is next to Shannon's car, but is not green.
3. Nancy's car is not green or yellow.
4. Dominick's car is between the blue and the brown car.
5. What color car belongs to Rosa? _____

green   red   blue   yellow   brown

Chapter 5 • 0 — 1000

Chapter 6
Fractions

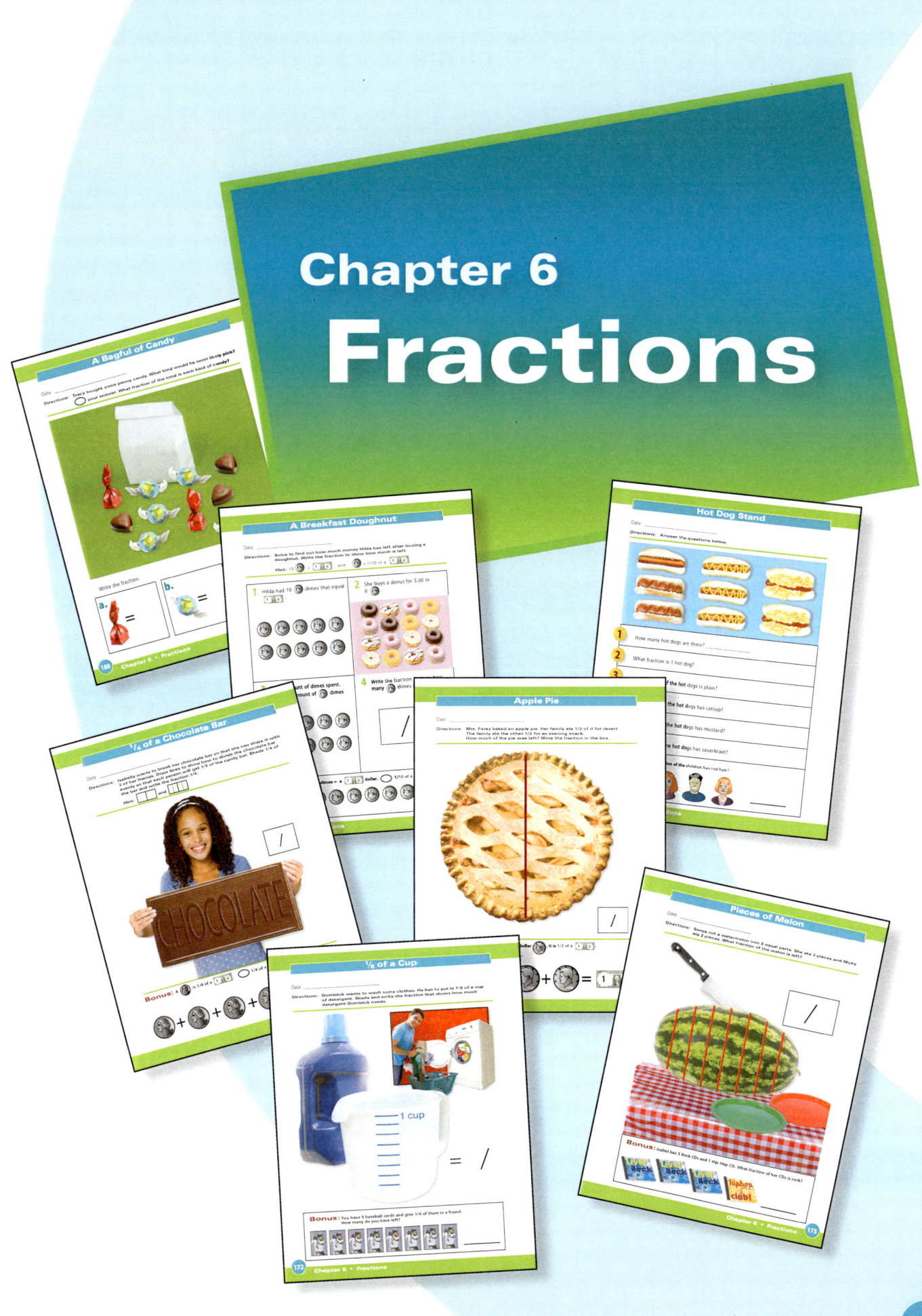

Chapter 6 ● Fractions

Lesson 1

Objective

S. will identify one-half of a whole.

Materials

- geometric shapes on board
- dry board
- markers
- pencils,
- student worksheet, p. 158: Football Snack

Procedure

1. Point to a shape and say that sometimes we need to divide a whole into parts.
2. The part of the whole we divide is called a **fraction.**
3. Draw a line through the shape to divide it into two equal pieces.
4. Shade one part. Say: "We divided this shape into two equal parts. This fraction is called **one-half.**"
5. Write the fraction $^1/_2$.
6. Point to the numerator.
7. The top number of the fraction is called a **numerator.** It tells how many parts or pieces are used.
8. Point to the bottom of the fraction.
9. The bottom part of the fraction is called a **denominator.**
10. It tells the total number of parts of the whole that have been divided.
11. Divide the other shapes in half.
12. Ask student volunteers to shade $^1/_2$ of the shape and to write the fraction $^1/_2$ beside it.
13. Continue until the students are firm.

Assign the worksheet: **Football Snack**

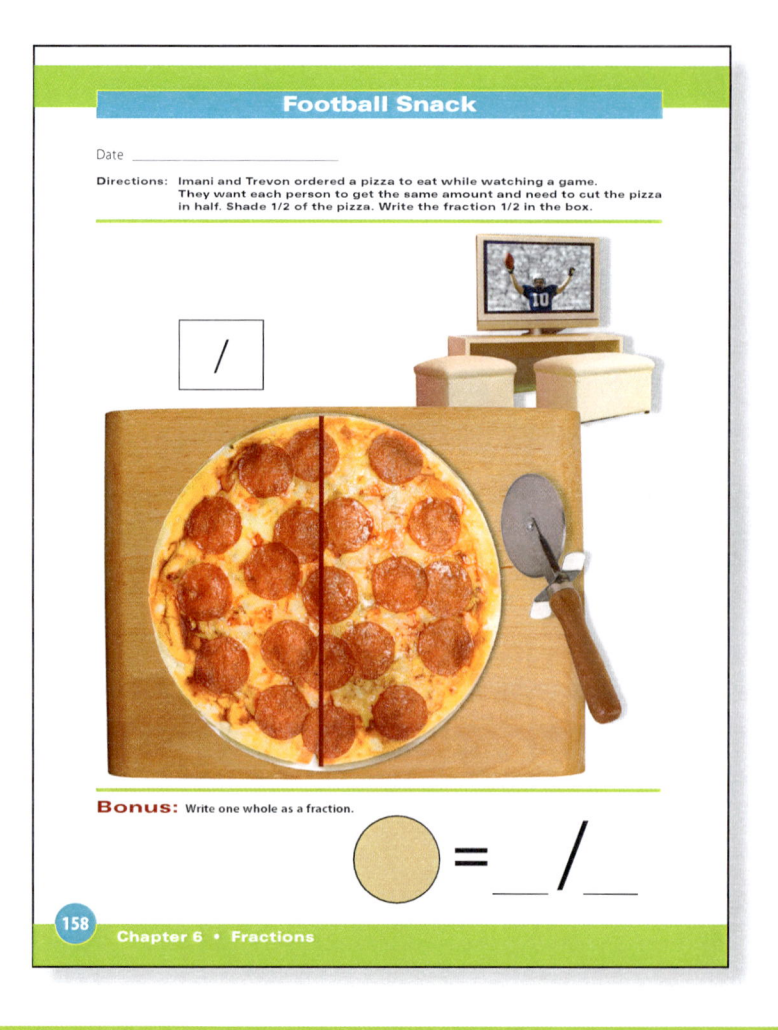

Lesson 2

Objective

S. will identify one-half of a set.

Materials

- sets of manipulatives, coins, or other small objects
- dry board
- markers
- pencils
- student worksheet, p. 159: Game Time

Procedure

1. Review how to identify **one-half** of a whole.
2. Review how to write the fraction $1/2$.
3. Tell students that sometimes we divide objects or groups of people into halves, e.g., making class teams, dividing bags of candy, money, etc.
4. Point to a set.
5. Demonstrate how to divide the set into two equal parts.
6. Point to a set. Say: "There are 6 nickels in this set. The nickels need to be divided into two equal parts or one-half of the set." Make two piles as the nickels are divided. Place 1 nickel in the first pile and 1 nickel in the second pile. Continue until all of the nickels have been used. Count the nickels in each pile. Say, "3 nickels are one-half of 6 nickels."
7. Repeat using another set of objects.
8. Ask individual students to divide the sets.
9. Continue until students are firm.

Assign the worksheet: **Game Time**

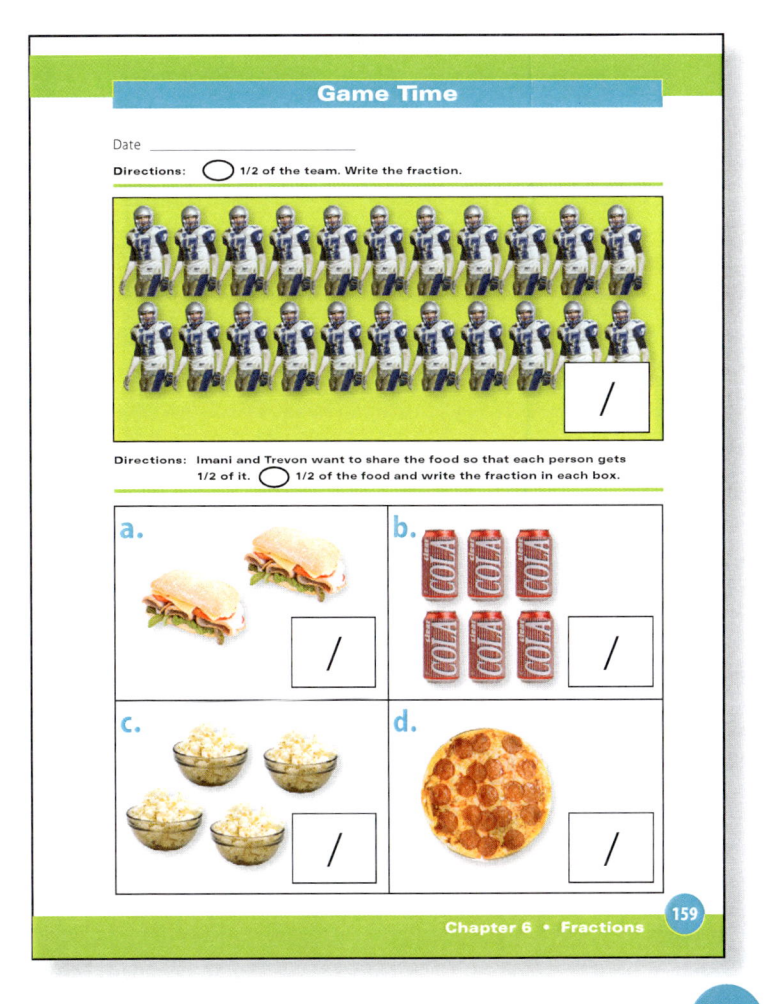

Chapter 6 • Fractions

Lesson 3

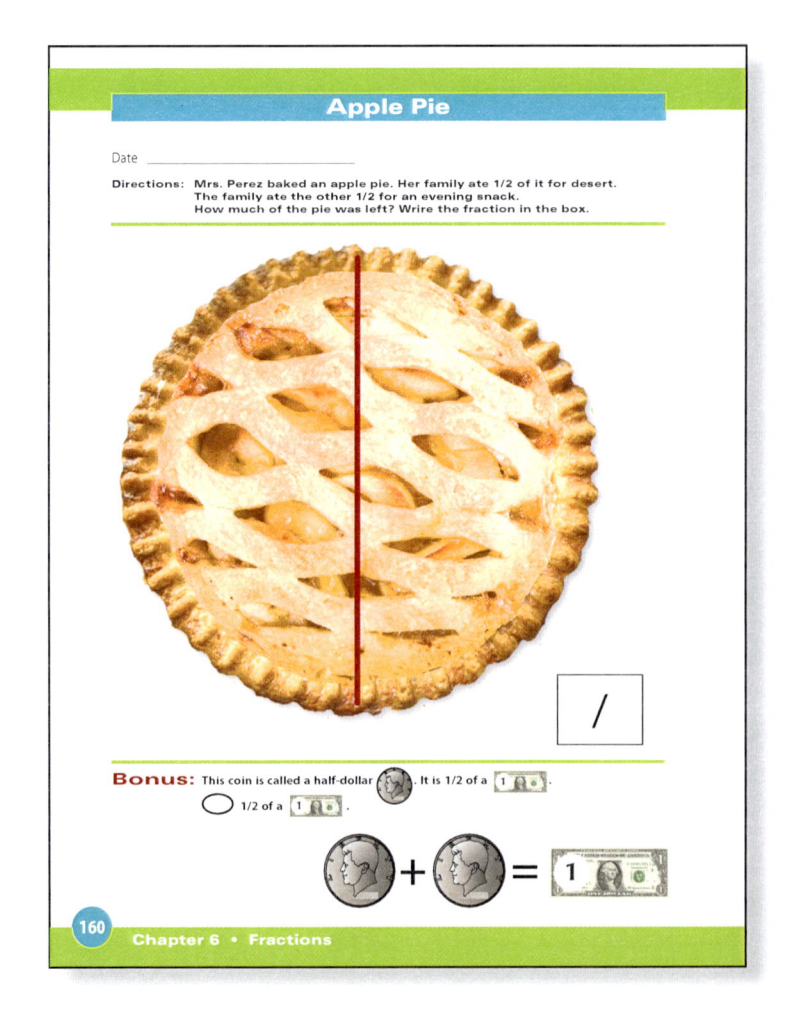

Objective

S. will solve a word problem using the fraction one-half.

Materials

- real-life objects that are divided into two parts, e.g., glasses of water, sandwiches, candy bars, etc.
- teacher-made problems
- dry board
- pencils
- student worksheet, p. 160: Apple Pie

Procedure

1. Create math problems using the real objects, e.g., Tamara ate one-half of her sandwich. She put the rest back in her lunch box. How much of her sandwich was left?

2. Ask individual students to solve additional problem using the other objects,

3. Continue until students are firm.

Assign the worksheet:
Apple Pie

Lesson 4

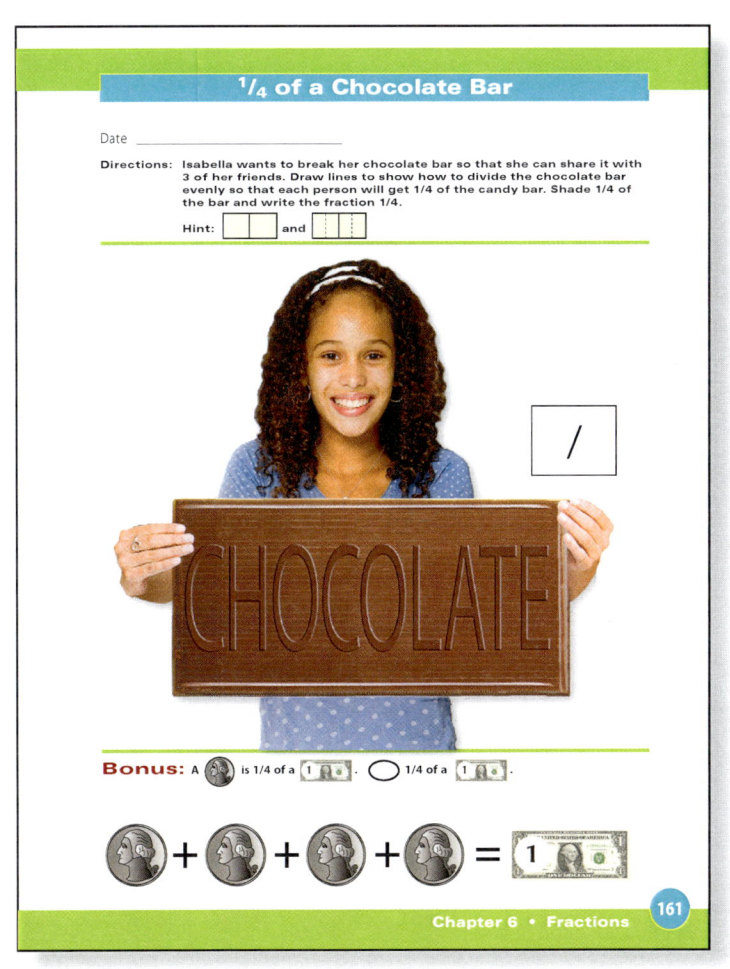

Objective

S. will identify one-fourth of a whole.

Materials

- geometric shapes on board
- dry board
- markers
- pencils
- student worksheet, p. 161: $1/4$ of a Chocolate Bar

Procedure

1. Review the fraction **one-half.**

2. Tell the students that they will learn a new fraction, **one-fourth.**

3. Point to a shape and tell the students this is how something can be divided into four equal parts, or fourths.

4. Divide the object in half and then into fourths.

5. Shade a fourth.

6. Point to the fraction and write **1/4** on the board next to it.

7. Ask individual students to divide the other objects into fourths and write the fraction **1/4.**

8. Continue until students are firm.

Assign the worksheet: $1/4$ of a Chocolate Bar

Lesson 5

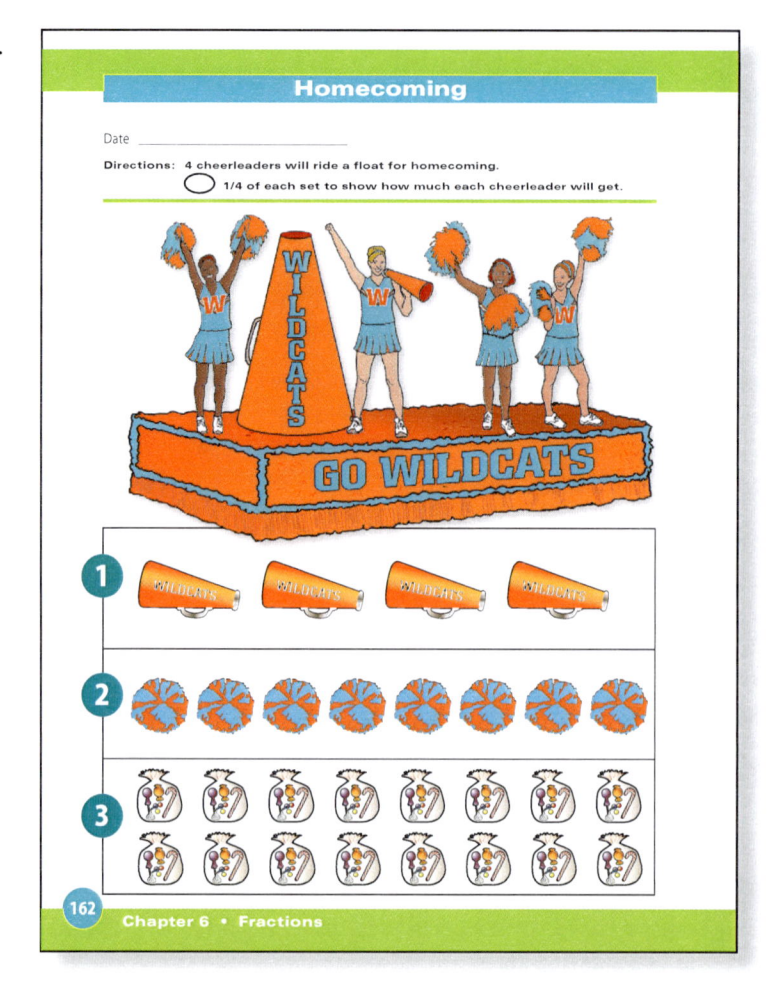

Objective

S. will identify one fourth of a set.

Materials

- set of manipulatives, coins, or small objects
- dry board
- markers
- pencils
- student worksheet, p. 162: Homecoming

Procedure

1. Review how to identify **one-fourth** of a whole.

2. Review how to write the fraction **1/4.**

3. Tell students that sometimes we divide objects or groups of people into fourths.

4. Brainstorm and list examples.

5. Point to a set, e.g., 16 pieces of candy.

6. Create a problem using the pieces.

7. Count out the pieces into four groups to find out how much is one-fourth of 16.

8. Continue with the rest of the objects until students are firm.

Assign the worksheet:
Homecoming

Lesson 6

Objective

S. will solve a word problem using fractions of different denominations.

Materials

- real-life objects (coins, etc.) that can be divided into fourths
- dry board
- markers
- student worksheet, p. 163: Homemade Cookies

Procedure

1. Review the fractions **one-half** and **one-fourth** with students.

2. Create problems where students must solve a problem using both fractions.

3. Lay out 8 quarters.

4. Say, "Samantha had 8 quarters. She used one-half of the quarters to pay for parking. She used one-fourth of the quarters to buy a candy bar. What fraction of the quarters did Samantha have left? How many quarters equal that fraction?"

5. Solve the problem by identifying one-half, then one-fourth, of the quarters.

6. Ask students what fraction is left.

7. Ask students how many quarters equal that fraction.

8. Continue until students are firm.

Assign the worksheet: **Homemade Cookies**

Lesson 7

Objective

S. will identify one-third of a whole.

Materials

- geometric shapes on board
- dry board
- pencils
- student worksheet, p. 164: More Pie

Procedure

1. Review previously learned fractions using the shapes on the board.
2. Introduce the fraction **one-third.**
3. Divide one of the geometric shapes into thirds.
4. Shade one-third of a shape and write the fraction $^1/_3$ next to it.
5. Repeat procedure.
6. Call on individual students to shade and write $^1/_3$.
7. Create problems which students must solve using thirds.
8. Continue until students are firm.

Assign the worksheet: **More Pie**

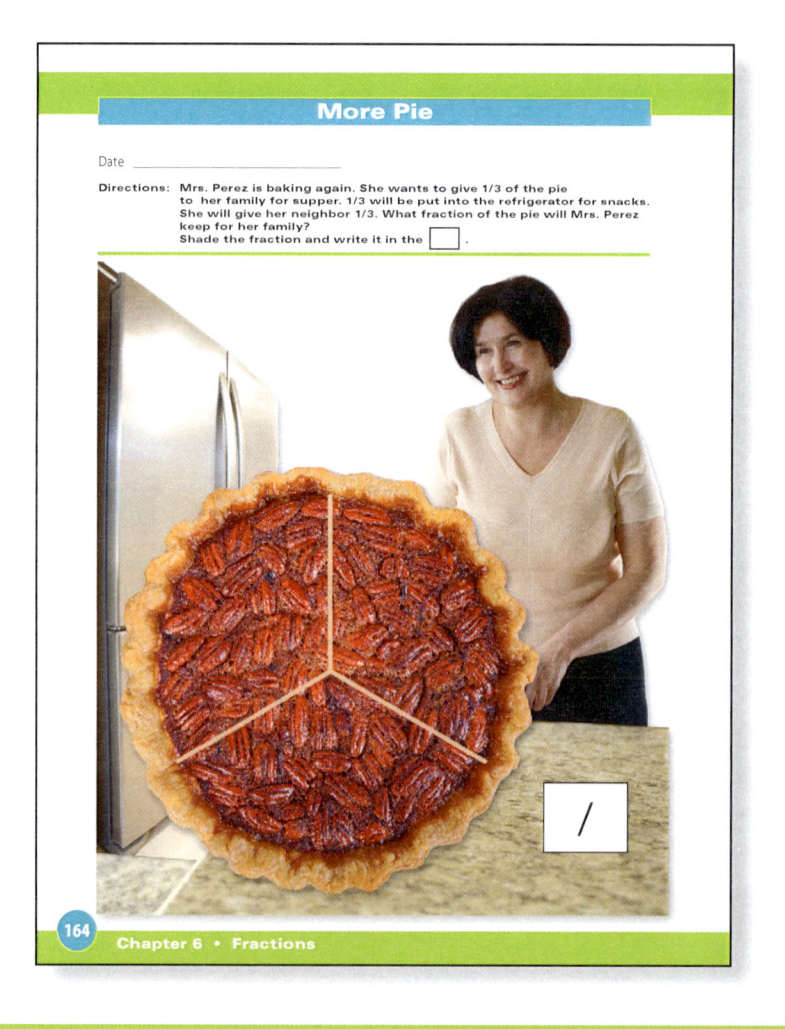

Lesson 8

Objective

S. will identify one-third of a set.

Materials

- sets of manipulatives or real items
- dry board
- pencils
- student worksheet, p. 165: Softball Practice

Procedure

1. Review how to identify **one-third** of a whole.

2. Review how to write the fraction $^1/_3$.

3. Tell students that sometimes we divide people or objects into thirds.

4. Pick a set of manipultives.

5. Count the number of objects, e.g., "There are 15 pencils in this set. How many pencils equal one-third of the set?" Next select three students and give each one an equal number until all of the pencils have been used.

6. Ask students to count the pencils they have.

7. Say, "Each student has 5 pencils. One-third of 15 is 5."

8. Continue with other sets until students are firm.

Assign the worksheet:
Softball Practice

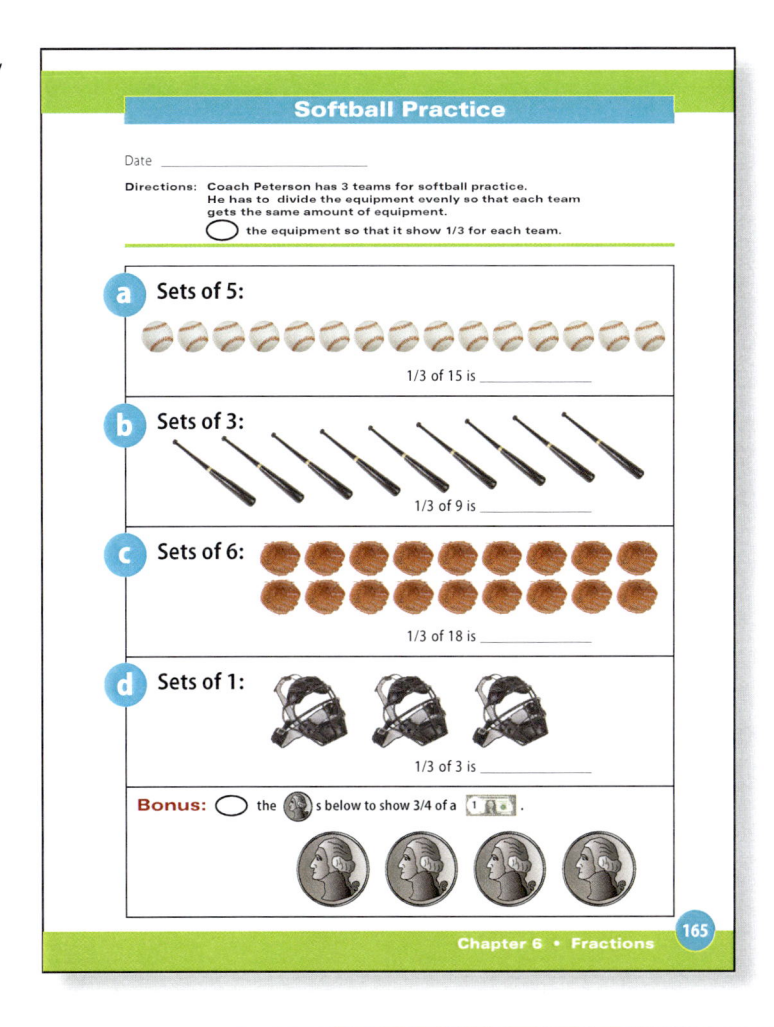

Lesson 9

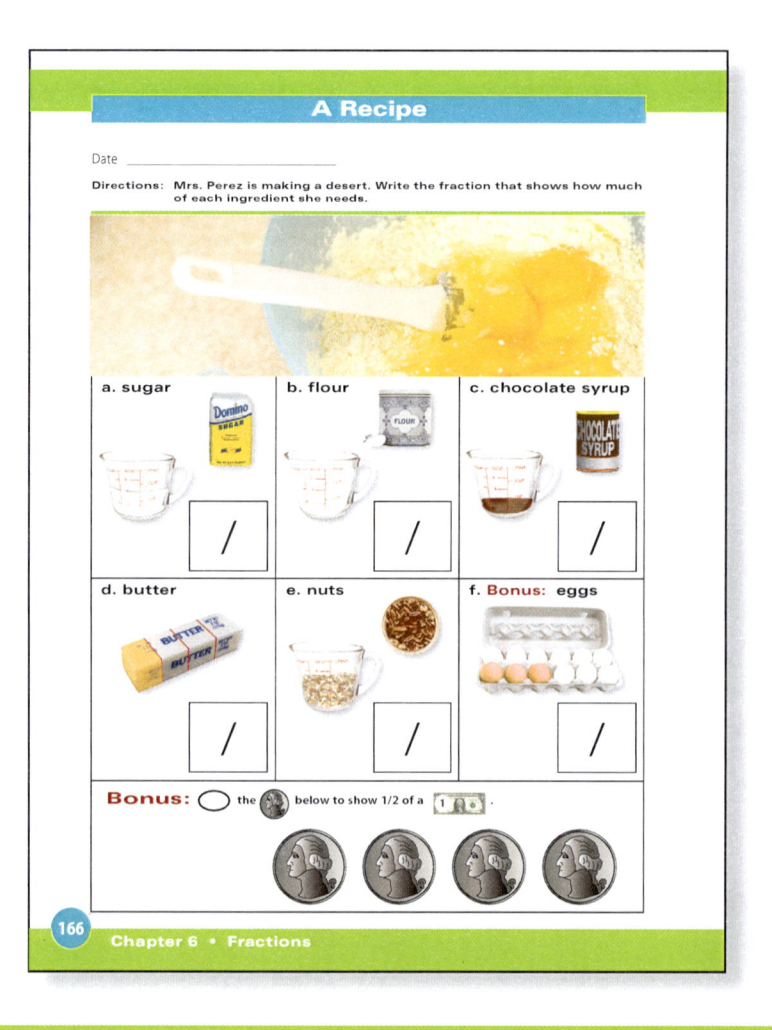

Objective

S. will identify and write a fraction.

Materials

- measuring cups
- sand, or other materials which can be used to fill a cup
- pencils
- student worksheet, p. 166: A Recipe

Procedure

1. Tell students that sometimes more than one kind of fraction has to be used in a recipe.

2. Demonstrate how to read the learned fractions using a measuring cup, e.g., "Select a cup. Pour sand into the cup, filling it to the one-half mark. Read the measurement and write the fraction $1/2$ on the board."

3. Repeat using a different fraction.

4. Call on individual students to fill the cup, e.g., two-thirds full.

5. After a student fills the cup and checks to be sure that the fraction is correct, he writes the measurement on the board.

6. Continue using all learned fractions until the students are firm.

Assign the worksheet:
A Recipe

Lesson 10

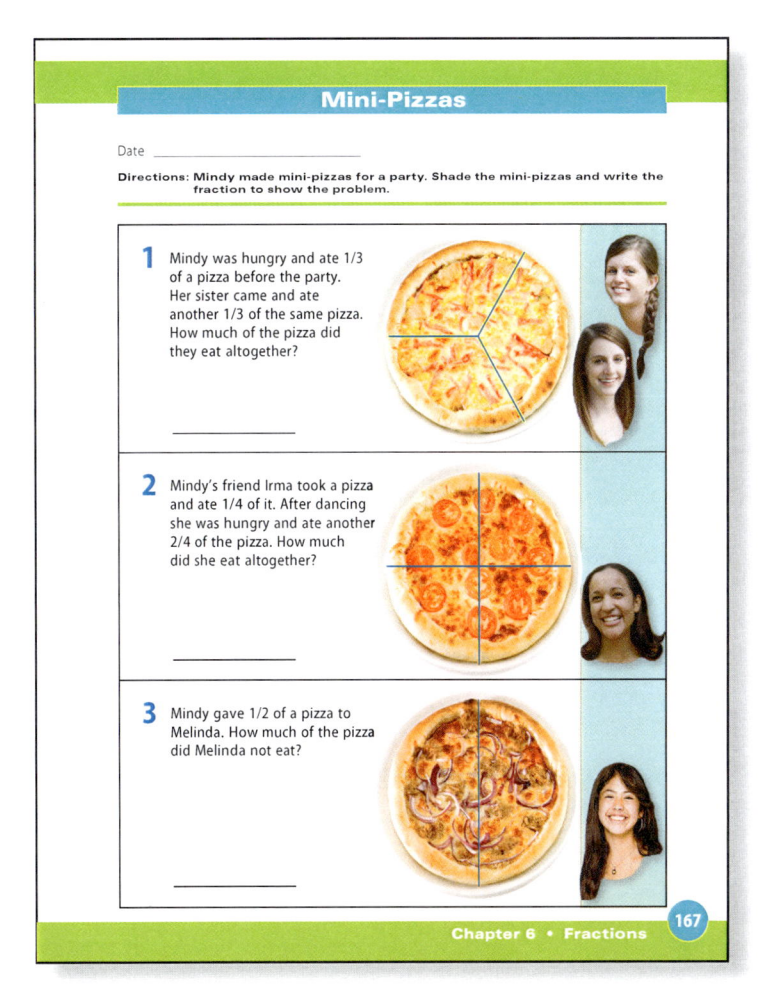

Objective

S. will solve problems using fractions of different denominations.

Materials

- circles on board divided into different fractions
- dry board
- markers
- pencils
- student worksheet, p. 167: Mini-Pizzas

Procedure

1. Point to one of the circles and create a problem using that circle, e.g., divide it into fourths. Say: "Isabel has an apple pie. She gave one-fourth to her brother. Shade one-fourth. She gave another one-fourth of the pie to her friend Erica. Shade that quarter. How much of the pie does Isabel have left?"

2. Point to the part of the pie that is not shaded and write the fraction to show how much of the pie is left.

3. Isabel has two-fourths of the pie left.

4. Encourage student to reduce the fraction if possible.

5. Repeat with other problems until all of the circles have been used.

Assign the worksheet:
Mini-Pizzas

Chapter 6 • Fractions

Lesson 11

Objective

S. will identify one-fifth of a whole.

Materials

- geometric shapes on board
- dry board
- markers
- pencils
- student worksheet, p. 168: One-fifth

Procedure

1. Review previously learned fractions using the shapes on the board.
2. Introduce the fraction: **one-fifth.**
3. Divide a shape into five equal parts.
4. Shade one-fifth of a shape and write the fraction on the board.
5. Repeat procedure.
6. Call on individual students to shade and write ¹/₅.
7. Create problems that students must solve using fifths.
8. Continue until firm.

Assign the worksheet: **One-fifth**

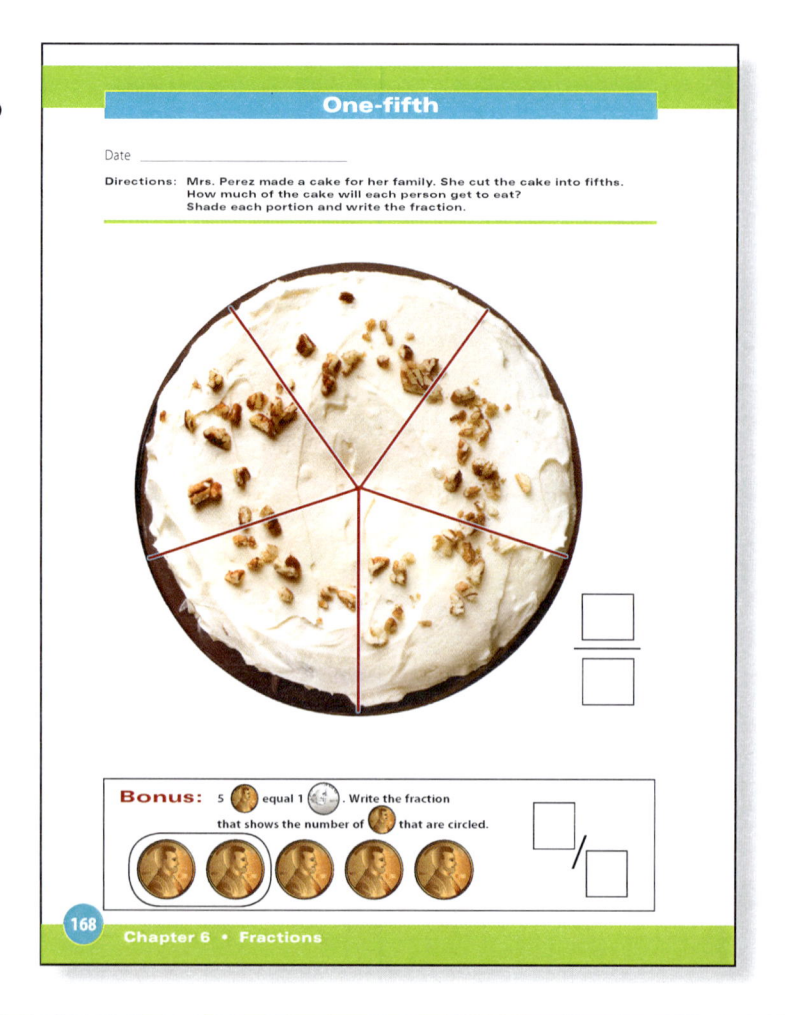

Lesson 12

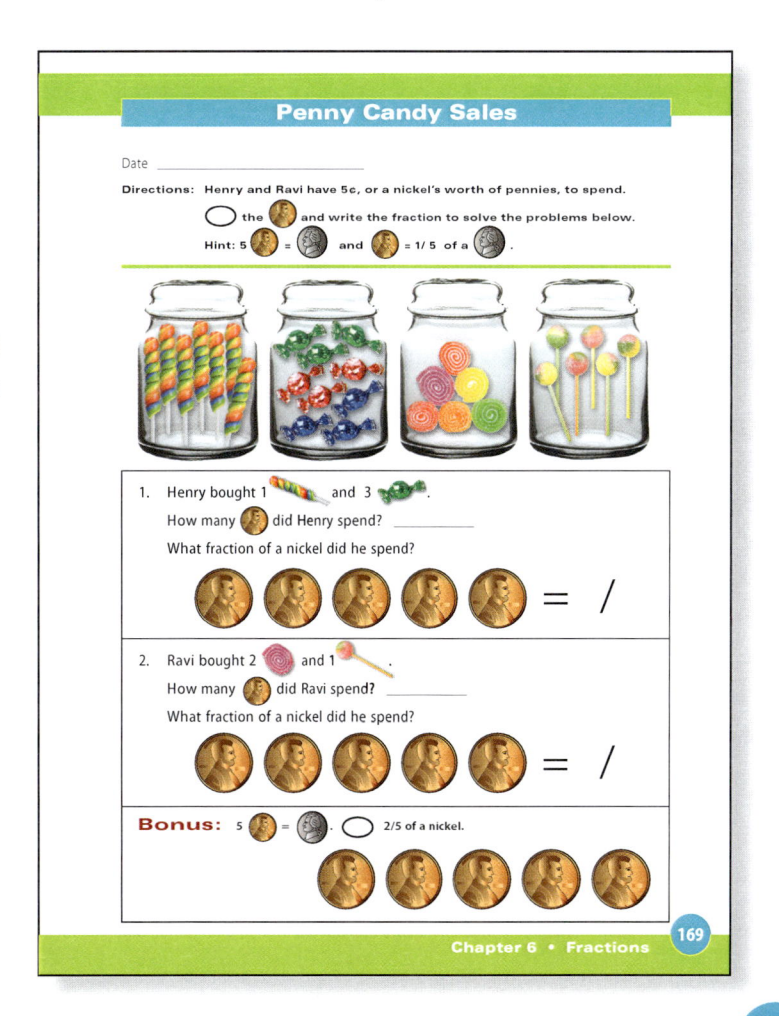

Objective

S. will identify the fractional part of a set, using fifths.

Materials

- real-life objects that can be used to identify the fifths of a set, e.g., coins, candy, etc.
- teacher-created problems
- dry board
- markers
- pencils
- student worksheet, p. 169: Penny Candy Sale

Procedure

1. Lay 5 nickels in a row in front of the students.

2. Count the nickels out loud. Say, "Five nickels equal 25 cents, or one quarter. One nickel is one-fifth of a quarter, two nickels are two-fifths of a quarter," etc.

3. Create problems where students must spend a fractional part of the quarter using nickels. Students must solve what fraction was spent and how many nickels equal that fraction, e.g., Harriet had 5 nickels. She spent 1 nickel on an eraser and 3 nickels on a pencil. What fraction of the total amount of money did Harriet spend?

4. Count out one nickel; count out 3 nickels and say: "Harriet spent $^4/_5$ of her money; $^4/_5$ equals 4 nickels."

5. Repeat using other problems and continue until the students are firm.

Assign the worksheet:
Penny Candy Sale

Chapter 6 • Fractions

Lesson 13

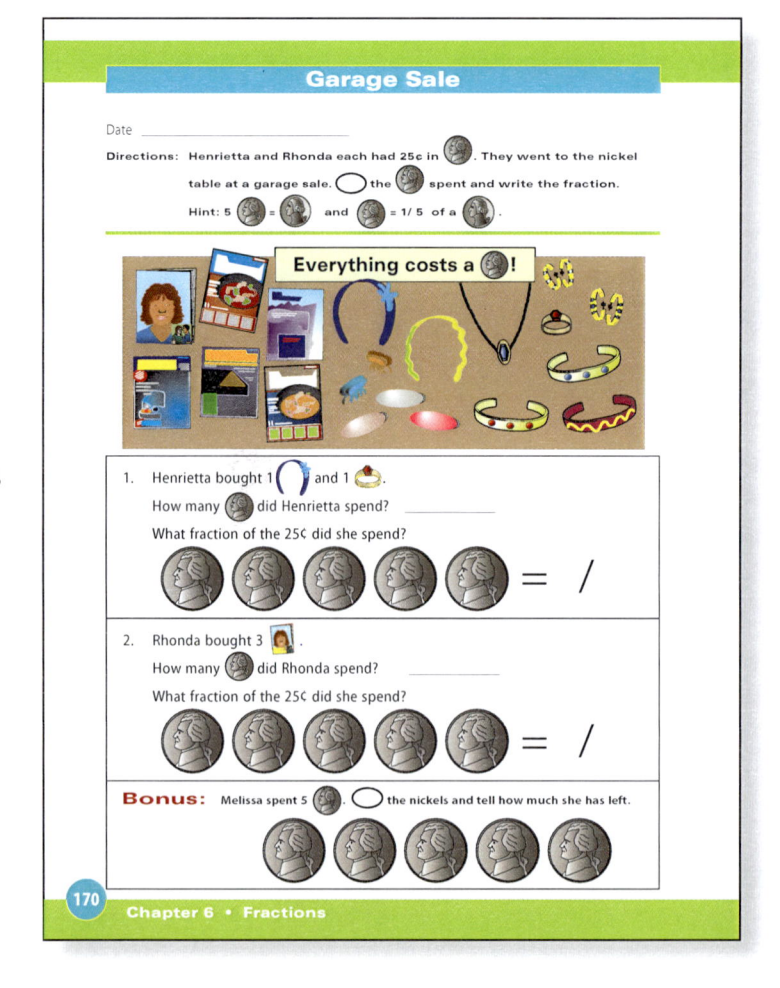

Objective

S. will identify the fractional part of the set, using fifths.

Materials

- sets of five nickels
- teacher-created problems
- dry board
- markers
- pencils
- student worksheet, p. 170: Garage Sale

Procedure

1. Using a set of nickels, count out the nickels to 25 cents.

2. Say: "Five nickels equals 25 cents or one quarter. One nickel is one-fifth of a quarter; two nickels are two-fifths of a quarter," etc.

3. Create problems where students must spend a fractional amount of the 25 cents. For example, Aaron went to a garage sale and bought a used baseball for fifteen cents. Remove the nickels. What fraction of the total amount did he spend? How many nickels equal that fraction?

4. Repeat using other problems until the students are firm.

Assign the worksheet:
Garage Sale

Lesson 14

Objective

S. will identify the fractional part of a set using fifths.

Materials

- sets of five $20.00 bills
- teacher-created problems
- dry board
- markers
- pencils
- student worksheet, p. 171: Clothing Sale

Procedure

1. Using a set of bills, count out the bills to one hundred.

2. Say, "Five $20-dollar bills equal $100.00. One $20-dollar bill is one-fifth of a 100, two $20-dollar bills are two-fifths of a 100," etc.

3. Create problems where students must spend a fractional portion of the $100.00, e.g., Heilke bought a dress on sale for $20.00 and a purse for the same price. Remove the $20-dollar bills. What fraction of $100.00 did Heilke spend? How many $20-dollar bills equal that fraction?

4. Call on individual students to answer the question.

5. Repeat with another problem and continue until students are firm.

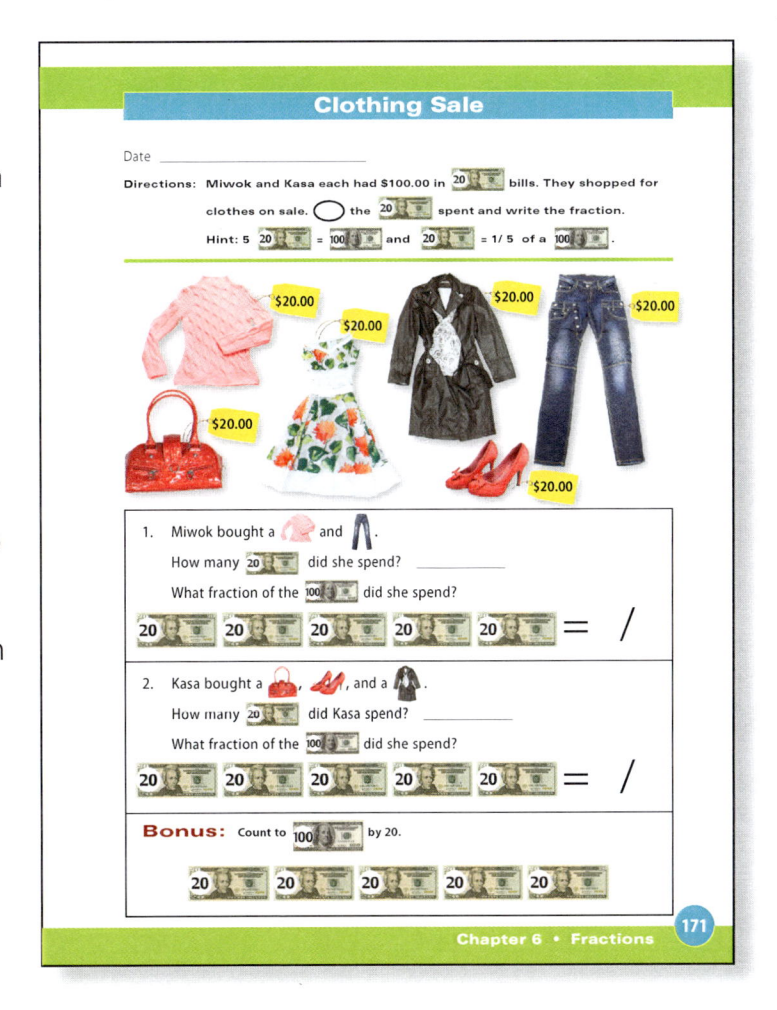

Assign the worksheet: **Clothing Sale**

Lesson 15

Objective

S. will identify one-eighth of a whole.

Materials

- measuring cups with eighths marked on them
- sand or other materials to be used to fill the cups
- dry board
- pencils
- student worksheet, p. 172: 1/8 of a Cup

Procedure

1. Review previously learned fractions using the cups.

2. Introduce **one-eighth.**

3. Point to the one-eighth mark and fill the measuring cup to that line.

4. Repeat.

5. Call on individual students to fill the cups to the one-eighth mark.

6. Continue until students are firm.

Assign the worksheet:
¹/₈ of a Cup

Explore Math Teacher's Manual

Lesson 16

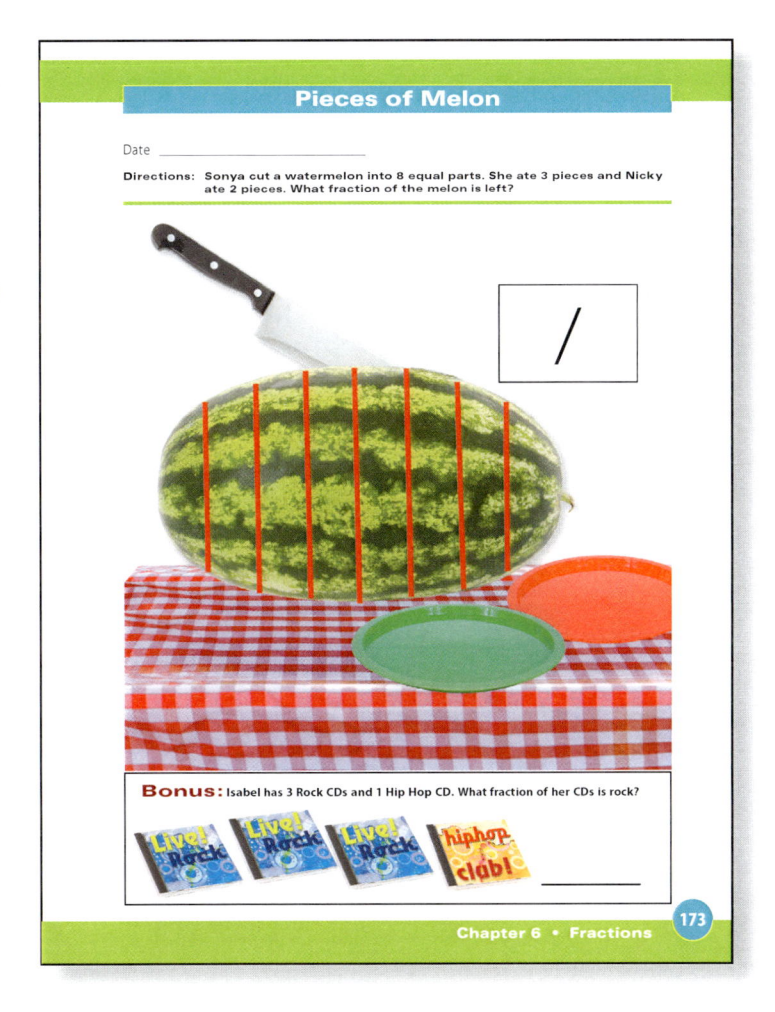

Objective

S. will solve a word problem using eighths.

Materials

- geometric shapes on board divided into eighths
- dry board
- markers
- pencils
- student worksheet, p. 173: Pieces of Melon

Procedure

1. Review the fraction **one-eighth** and how to write it ($^1/_8$).

2. Create problems where students must solve it using eighths, e.g., point to a shape and say: "Melinda made a cake. She divided it into eighths. Divide a rectangle into eight equal pieces. She gave one-eighth of the cake to her brother. Shade that part. She gave two pieces to her twin sister. Shade two more eighths. How many pieces of the cake does she have left? What fraction of the cake does she have left?"

3. Call on individual students to answer the questions.

4. Repeat using another shape and problem.

5. Continue until students are firm.

Assign the worksheet:
Pieces of Melon

Chapter 6 • Fractions

Lesson 17

Objective

S. will identify parts of a set using eighths.

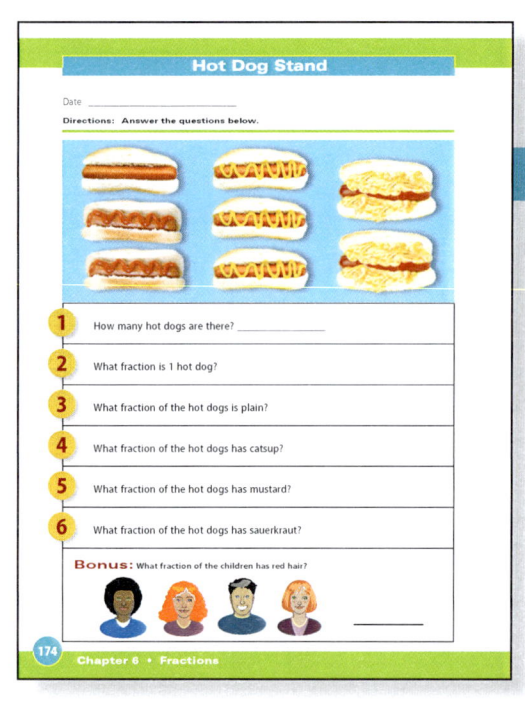

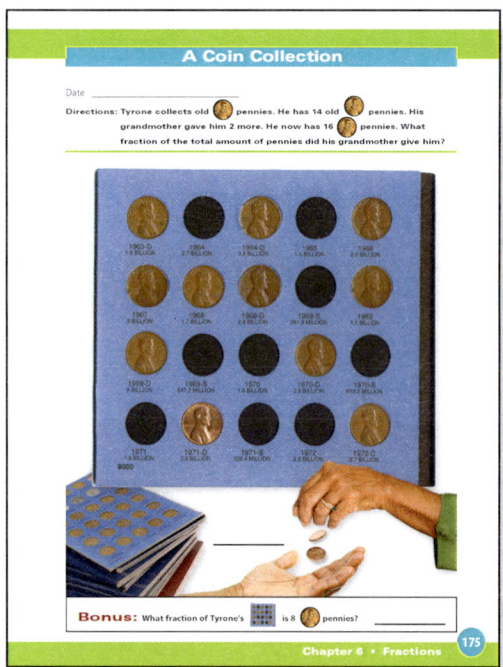

Materials

- sets of manipulatives or real-life objects that can be divided into eighths
- dry board
- markers
- pencils
- student worksheets, pp. 174–175: Hot Dog Stand and A Coin Collection

Procedure

1. Create a teacher-made problem using the manipulatives, e.g., count out a set of 8 pennies. Say, "Harold had 6 pennies in his coin collection." Count out 6 pennies and place them in a row. "His mother gave him 2 more pennies; now he has 8 pennies." Place the 2 pennies near the row of 6. "What fraction of the set of pennies did his mother give him?" Tell the student that one penny is one-eighth of the set. Two pennies are two-eighths of a set, etc. Ask the students how many eighths of a set are 2 pennies.

2. Repeat, using another problem and set of manipulatives, e.g., Tyrone cut a pie into 8 parts. His brother ate 2 pieces and Tyrone had 3. What fraction of the pie is left? Take away the parts of the set that were "eaten" and count the fraction that is left. Three-eights of the pie is left.

3. Continue until students are firm.

Assign the worksheets:
Hot Dog Stand and **A Coin Collection**

Lesson 18

Objective

S. will identify one-tenth of a whole.

Materials

- geometric shapes on the board divided into previously learned fractions and tenths
- dry board
- markers
- pencils
- student worksheet, p. 176: $^1/_{10}$

Procedure

1. Review previously learned fractions using the shapes on the board.

2. Introduce the fraction **one-tenth.**

3. Point to a tenth on one of the geometric shapes.

4. Shade the fraction and write the fraction $^1/_{10}$ to represent the shaded part.

5. Repeat procedure.

6. Create problems that students must solve using tenths.

7. Continue until students are firm.

Assign the worksheet: $^1/_{10}$

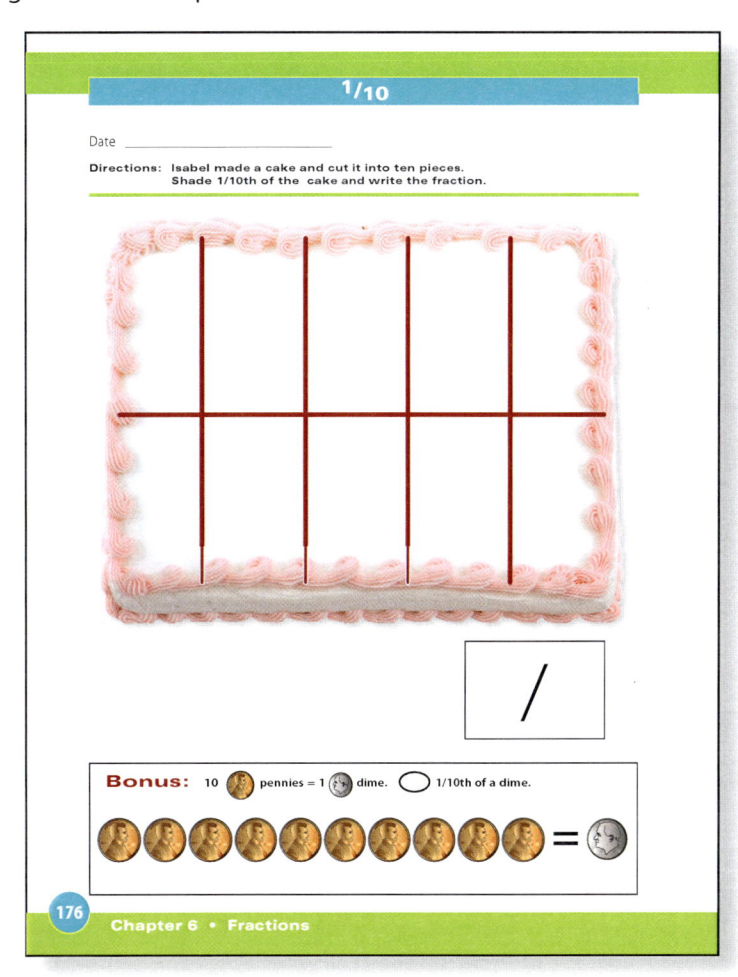

Lesson 19

Objective

S. will solve a word problem using tenths.

Materials

- sets of ten pennies, dimes and ten-dollar bills
- dry board
- markers
- pencils
- student worksheets, pp. 177-179: A Penny Collection, A Breakfast Donut and A New Cell Phone

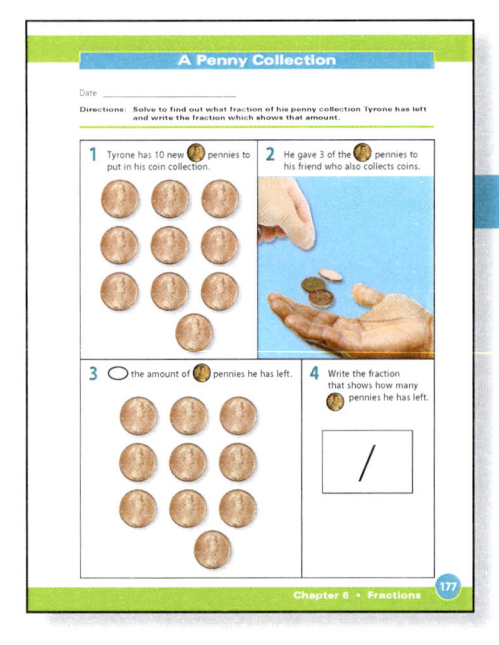

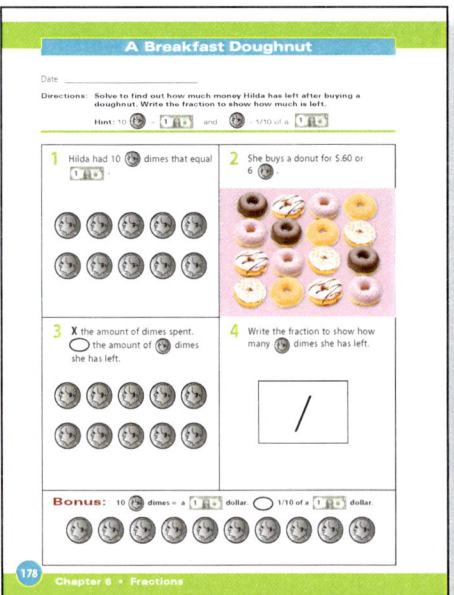

Procedure

1. Create problems where students must spend a fractional part of the money in each set of bills.

2. Count ten pennies out loud. Say, "10 pennies equals a dime. One penny is one-tenth of a dime, 2 pennies are two-tenths of a dime, etc." Continue until all of the pennies have been counted.

3. Create problems using the pennies, e.g., Carlos has 10 pennies. He buys a piece of gum with 6 of the pennies. How many pennies does he have left? What fraction of the original amount does he have left?

4. Count out 6 pennies. Count the pennies that are left. Carlos has 4 pennies left, or 4/10 of the set.

5. Repeat a similar procedure using dimes and ten-dollar bills.

6. Continue until students are firm.

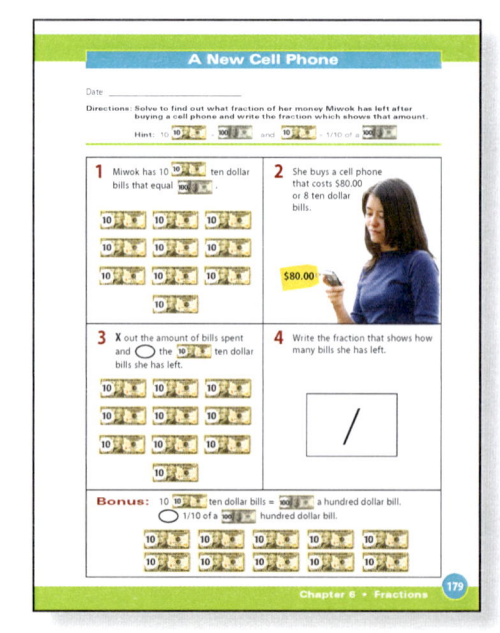

Assign the worksheets:
A Penny Collection
A Breakfast Doughnut
A New Cell Phone

Explore Math Teacher's Manual

Lesson 20

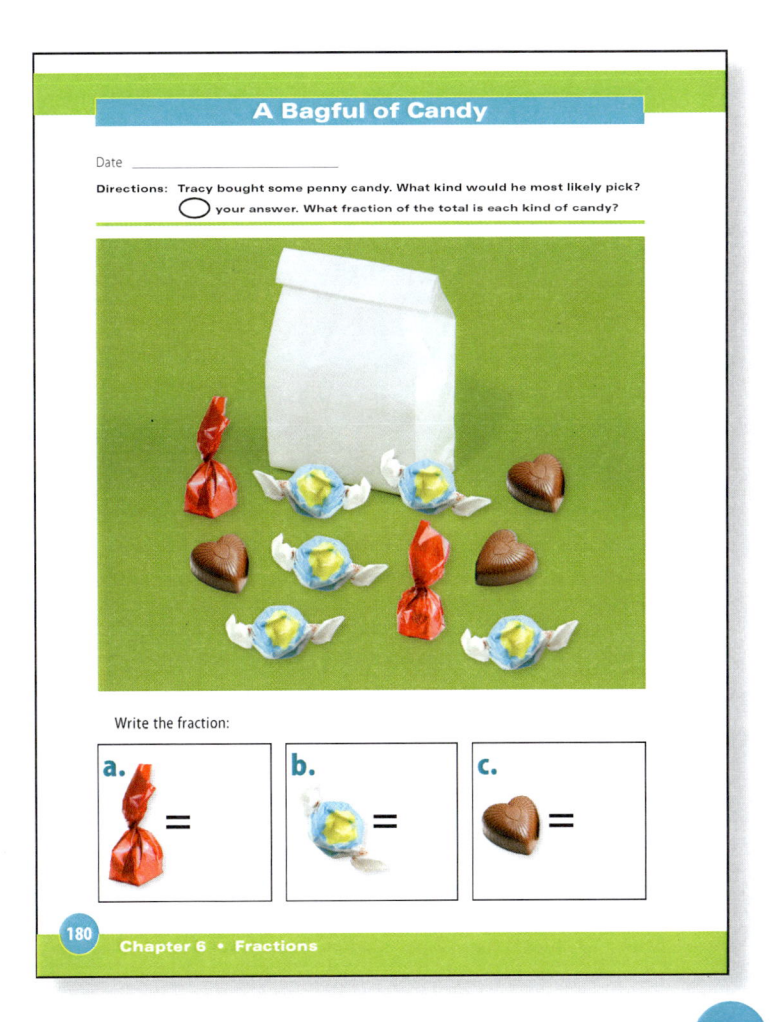

Objective

S. will solve a math riddle using fractions.

Materials

- small items of various kinds and shapes in plastic bags
- dry board
- markers
- pencils
- student worksheet, p. 180: A Bagful of Candy

Procedure

1. Select a plastic bag.

2. Tell students to look at the bag.

3. Explain that there are different amounts of each item in the bag.

4. Students are to predict which item would be picked most often, if the items were pulled from the bag one at a time.

5. Write the predictions on the board.

6. Take the items from the bag one at a time and place them in separate piles according to shape, kind, etc.

7. Count each pile and write the total amount for each item.

8. Determine the fractional part of the whole set for each shape or kind.

9. Compare the total to the predictions.

10. Discuss results.

Assign the worksheet:
A Bagful of Candy

Chapter 6 • Fractions

Lesson 21

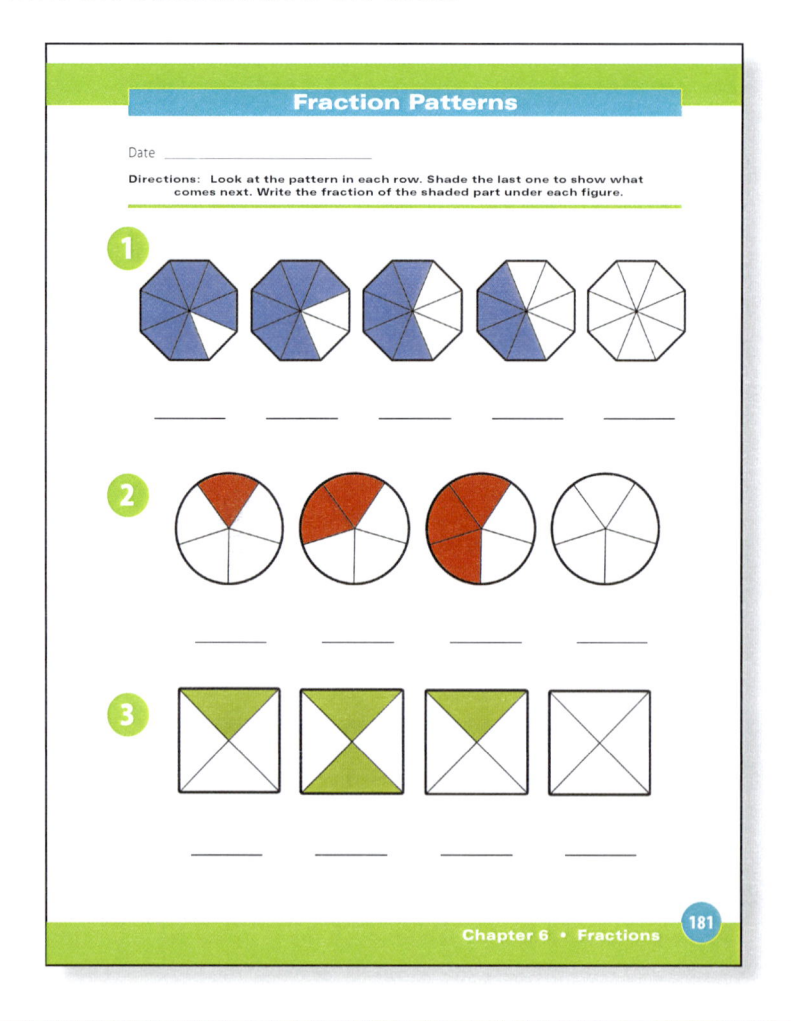

Objective

S. will complete fraction patterns

Materials

- circles drawn on the board, using a similar pattern, as shown on the Fraction Patterns worksheet
- dry board
- markers
- student worksheet, p. 181: Fraction Patterns

Procedure

1. Point to a fraction pattern.

2. Count out how many parts the circles were divided into, to create the pattern.

3. Count each shaded part and write the fraction below the circle.

4. Continue until the last circle is reached.

5. Determine the pattern that was used and complete the pattern using the last circle.

6. Repeat the procedure using the next pattern.

7. Continue until the students are firm.

Assign the worksheet:
Fraction Patterns

Lesson 22

Objective

S. will compare the size of different fractions.

Materials

- Fractions from student worksheets, pp. 182-183: More Fractions 1 and 2—printed, cut out, and laminated

Procedure

1. Select two of the learned fractions.

2. Demonstrate how to compare the size of the fractions, e.g., one-half and one-fourth. Place the one-half fractions in front of the students. Ask them to determine or predict whether the fraction one-fourth is bigger or smaller than one-half. Place the fraction one-fourth on top of the one-half fraction. Compare the fractions. Next, determine how many fourths will equal a half, a whole, etc.

3. Repeat, using another set of fractions.

4. Continue until students are firm.

Assign learned fractions using the worksheets: **More Fractions 1** and **2**

Note: These fraction pieces can also be used to teach how to reduce fractions.

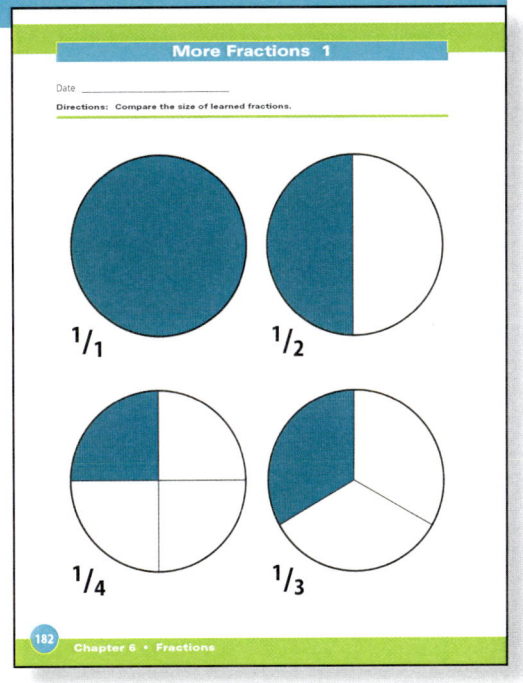

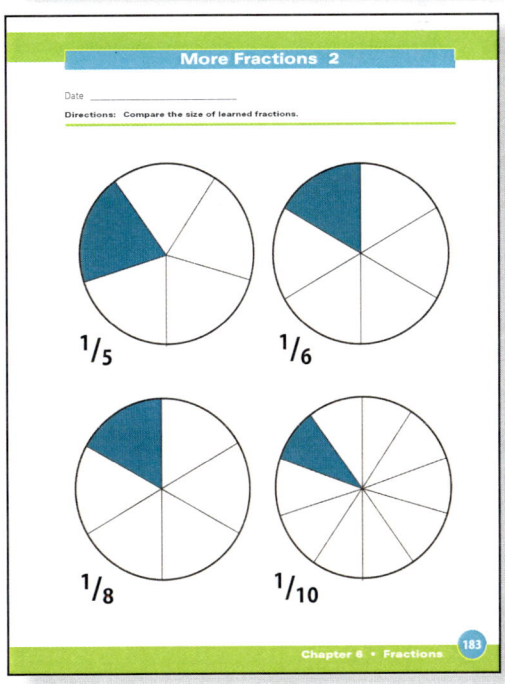

Chapter 7
Answer Key

Chapter 1 • Vocabulary

p. 12

1. ◯ the first number.
   ②, 4, 6, 8, 10, 12
2. ◯ the number that is in between.
   10, ⑪, 12
3. ◯ the last number.
   2, 3, 4, 5, ⑥
4. ◯ the person that is behind. 
5. ◯ the person in front of the line. 
6. ◯ the last person. 
7. ◯ the boy that is between 2 girls.

p. 13

1. total amount $3 + 5 = 8$
2. total score $2 + 1 = 3$ points (pts.)
3. in all $4 + 3 = 7$
4. altogether $\$6.00 + \$5.00 = \$11.00$

p. 14

1. difference $12 - 9 = 3$ points (pts.)
2. left $7 - 5 = 2$ 
3. left $8 - 3 = 5$ 
4. left $\$6.00 - \$4.00 = \$2.00$

p. 15

1. subtract $12 - 3 = 9$ 
2. add $6 + 2 = 8$ 
3. add $8 + 2 = 10$ 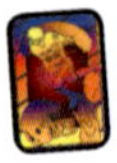
4. subtract   12 points − 9 points = 3 points

p. 16

1. add $4 + 6 = 10$ 
2. subtract $17 - 8 = 9$ 
3. subtract $6 - 4 = 2$ 
4. add $10 + 4 = 14$

Chapter 2 • 0–12

p. 18

Bonus:

p. 19

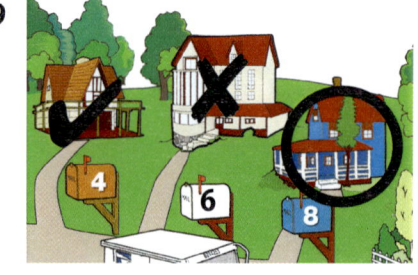

p. 20

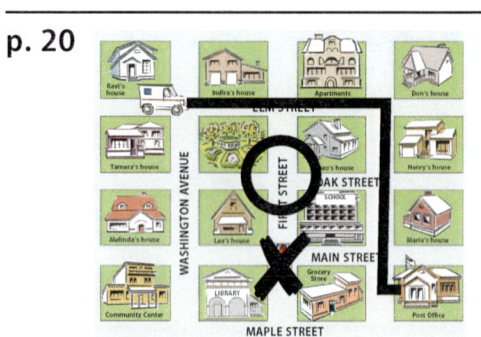

1. 5 blocks   2. 3 blocks

p. 21

$5 + 6 = 11$ letters   + (add)

144

Explore Math Teacher's Manual

p. 22

6 + 4 = **10 boxes**

Bonus: 4 boxes

p. 23

8 + 4 = **12 baseballs**

p. 24

3 + 4 = **7 pictures**

p. 25

Bulls 0 + 7 = **7 pts.**

Colts 3 + 6 = **9 pts.**

p. 27

1. Balls 3 + **7** = 10
2. Bats 2 + **3** = 5
3. Gloves 3 + **4** = 7
4. Catcher's masks 1 + **0** = 1
5. Hard caps 6 + **6** = 12
6. answers vary

p. 28

5 pts + 3 pts + 2 pts. = **10 pts.**

p. 29

3 + 2 + **3** = 8

p. 30

1. ($3.00) + $2.00 + $1.00 = $6.00

p. 31

9¢ − 7¢ = **2¢**

p. 32

$9.00 − $6.00 = **$3.00**

p. 33

2 + 2 + 2 + = 6   12 − 6 = **6**

p. 35

1.
2.
3.

4. Answers vary

p. 36

1.   2.

p. 37

1.   2.

Bonus: 7 hours

p. 38

1.   2.

p. 39

1. 1:00
2. 10:00
3. 4 hours
4. 4 hours
5. 1 hour

p. 41

1. 7 days
2. Sunday
3. Saturday
4. Tuesday
5.  (violin)
6.  (volleyball player)
7.  (football player)
8. 2 days
9. 3 days
10.  (box of chocolates)

Bonus: answers vary

p. 43

1. January
   February
   March
   April
   May
   June
   July
   August
   September
   October
   November
   December

2. February
3. June
4.  (basketball hoop)
5.  (star)
6. 6 months
7. 9 months
8. 6 months
9. 4 months

Bonus: answers vary

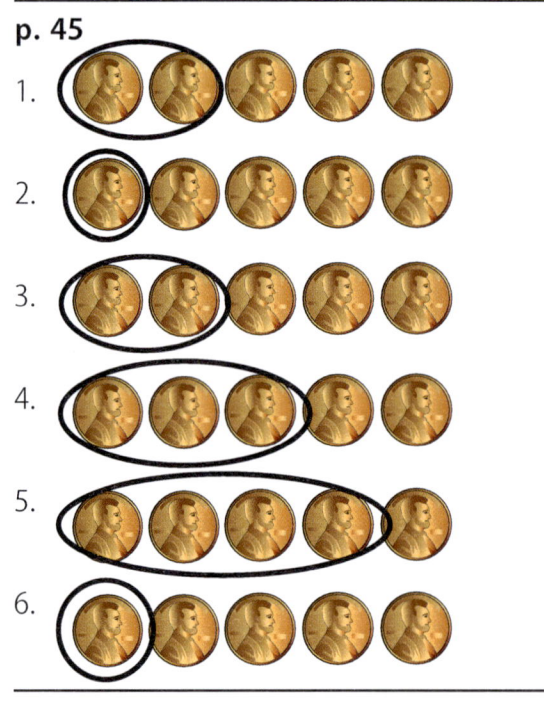

p. 44

1.  – 6.  (coin grouping exercises)

p. 45

1. – 6. (penny grouping exercises)

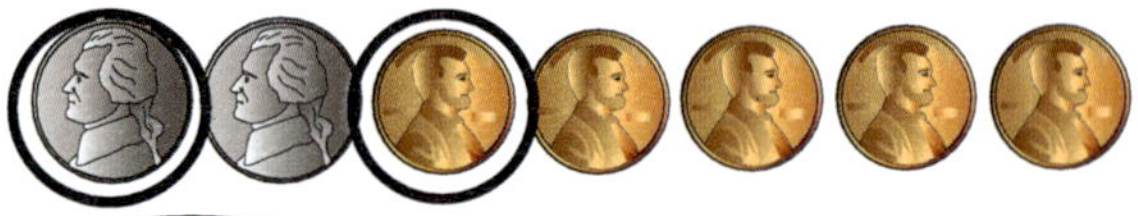

pp. 47

1.  (football)
2.  (music/CD)
3. 11 people
4. 8 people
5. 2 people
6. 6 people

pp. 49

1. 8 players
2. 4 players
3. 6 players
4. 4 players
5. 2 players
6. answers vary

p. 50

1. 4 tails
2. 6 ears
3. 12 legs
4. 10 eyes

p. 51

answers vary

Chapter 3 • 0–18

p. 54

6 + 5 = 11

p. 55

1. and 5.

2. 3 blocks
3. 5 blocks
4. 3 blocks
5. 3 blocks

Bonus: 14 blocks

p. 56

2 + 9 = 11

Bonus: 8 apples

p. 57

4 + 8 = 12

pp. 59

1. 5 + 10 = 15 lb.
2. 6 + 3 = 9 lb.
3. 10 + 2 = 12 lb.
4. 3 + 5 = 8 lb.

pp. 60

1.
2.
3.
4.

pp. 61

1. $5.00 + $2.00 = $7.00
2. $6.00 + $3.00 = $9.00
3. $4.00 + $2.00 = $6.00
4. $3.00 + $5.00 = $8.00

p. 62

HOME: 5 + **6** = 11 pts.

AWAY: 3 + **2** = 5 pts.

p. 63

13 + 5 = 18

p. 64

1. 4 + 1 + 8 = **13 pts.**
2. 3 + 1 + 2 = **6 pts.**
3. 10 + 1 + 4 = **15 pts.**
4. 3 + 1 + 1 + 4 = **9 pts.**
5. 10 + 1 + 1 + 1 + 1 = **14 pts.**

Bonus: Most points: Player 3
         Least points: Player 2

p. 65

Day One: 7 mi. + 4 mi. + 2 mi. = **13 mi.**

Day Two: 8 mi. + 3 mi. + 7 mi. = **18 mi.**

p. 66

4 + 3 + **6** = 13 free throws

p. 67

16 − 8 = **8** boxes

p. 68

17 − 13 = **4 pts.**

p. 69

18 − 9 = **9**

Bonus: warmer

pp. 71

1. 18° − 10° = **8°**
2. 16° − 8° = **8°**
3. 9° − 2° = **7°**
4. 12° − 7° = **5°**
5. 17° − 8° = **9°**
6. 17° − 11° = **6°**
7. 15° − 12° = **3°**

p. 72

Softball Toss: 15 − 6 = **9 bottles**

Dart Throw: 12 − 5 = **7 balloons**

p. 73

14 − 3 = 11 tickets

11 − 2 = **9 tickets**

p. 74

HOME: ⑦ + 7 = 14 pts.

AWAY: ③ + 14 = 17 pts.

p. 75

1.  (clock) 2.  (clock)

p. 76

1.  (clock) 2.  (clock)

p. 77

1.  (clock) 2.  (clock)

p. 78

1.  (clock) 2.  (clock)

3.  (clock) 4.  (clock)

p. 79

Saturday

Sunday

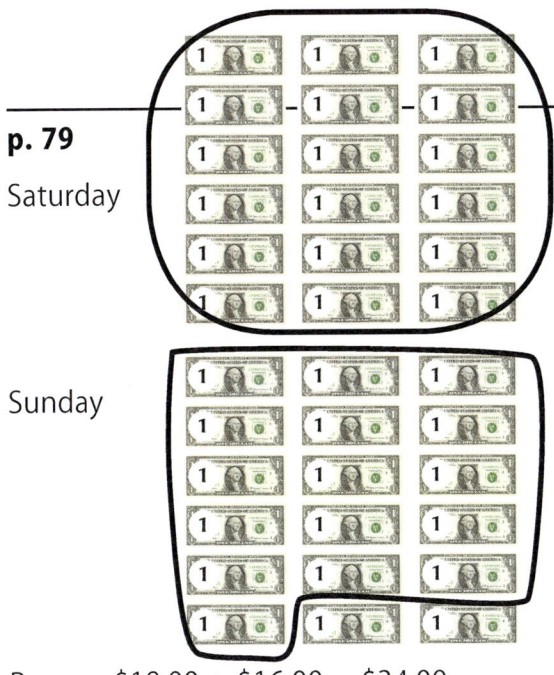

Bonus: $18.00 + $16.00 = $34.00

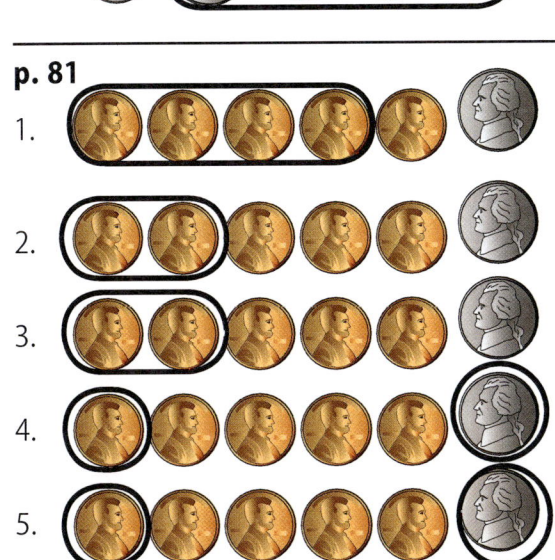

p. 80

1.
2.
3.
4.
5.
6.

p. 81

1.
2.
3.
4.
5.
6.

p. 83

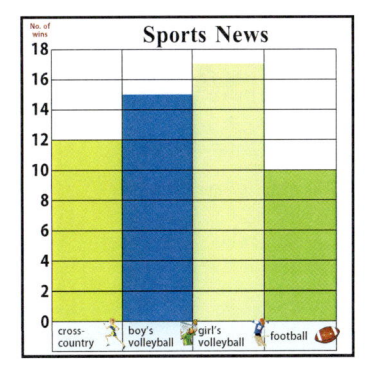

Bonus: 17 − 12 = 5 games

p. 84

	Inventory		
	Item	No. on the shelf	Total
1.	chicken soup	|||| ||||		11
2.	canned spaghetti	|||| |||| |||| |||	18
3.	tuna fish	|||| ||	7
4.	macaroni	|||| |||| ||||		16
5.	pizza mix	|||| |||| ||||	14

Bonus: 16 − 11 = 5 cans

p. 85

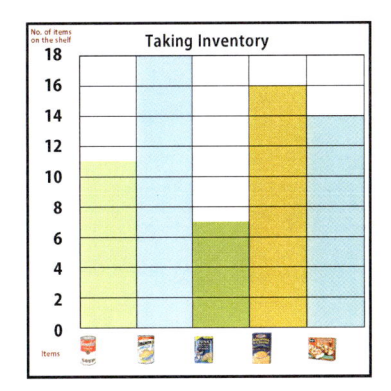

Bonus: 14 − 7 = 7 more pizza mixes

p. 86

1. 14
2. 18
3. 16

149

Chapter 7 ● Answer Key

p. 87

Color	Alex	Tony	Kareem
red	Y	N	N
blue	N	Y	N
black	N	N	Y

(Tony circled)

Event	Tamika	Alexandra	Isabel
hurdles	N	Y	N
pole vault	Y	N	N
broad jump	N	N	Y

(Isabel circled)

Chapter 4 • 0–100

p. 91

p. 92

1.
2.
3. 
4. 3 blocks
5. 6 blocks
6. 4 blocks

p. 93

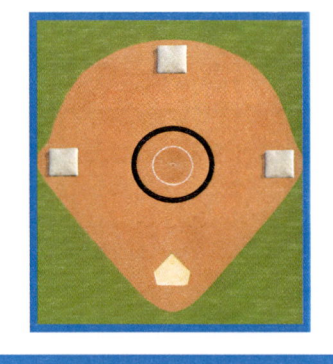

p. 95

1. 
2. 
3. 
4. 
5. 
6.

p. 97

1. 36 floors
2. 36 − 9 = **27 floors**
3. 24 − 1 = **23 floors**
4. 32 − 18 = **14 floors**
5. 29 − 25 = **4 floors**
6. 15 − 2 = **13 floors**

p. 98

32 + 21 = **53 newspapers and magazines**

p. 99

   43 roast beef sandwiches
+ 34 ham and cheese sandwiches
= **77 sandwiches**

Bonus: 3 sandwiches

p. 100

28 + **12** = 40 handouts

p. 101

Game 1:
   State: **43 pts.** + 25 pts. = 68 pts
   Away: **26 pts.** + 13 pts. = 39 pts.

Explore Math Teacher's Manual

Game 2:
    State: **43 pts.** + 40 pts. = 83 pts.
    Away: **50 pts.** + 22 pts. = 77 pts.

Game 3:
    State: **51 pts.** + 34 pts. = 85 pts.
    Away: **55 pts.** + 42 pts. = 97 pts.

Game 4:
    State: **31 pts.** + 36 pts. = 67 pts.
    Away: **22 pts.** + 44 pts. = 66 pts.

Bonus: Win 3   Lose 1

p. 103

1.

2. 77 − 62 = **15 students**

3. 88 − 44 = **44 students**

4. 77 − 23 = **54 students**

5. 23 + 62 = **85 students**

6. 26 − 23 = **3 students**

p. 104

85° − 60° = **25°**

p. 105

1.

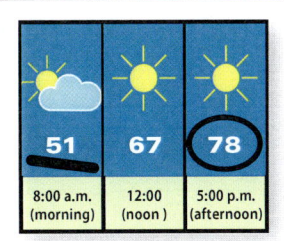

2. 67° − 51° = **16°**

3. 78° − 67° = **11°**

4. 78° − 51° = **27°**

p. 106

Sunday

p. 107

1. 100 − 90 = **10°** (degrees)

2. 99 − 83 = **16°**

3. 98 − 83 = **15°**

4. 84 − 70 = **14°**

5.

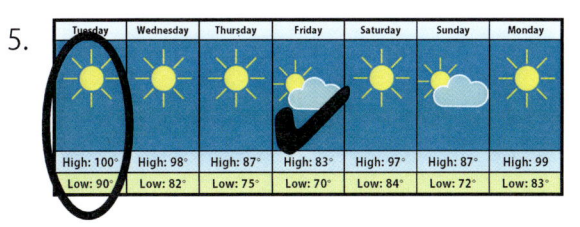

p. 108

$68.00 − $52.00 = $16.00

p. 109

$38.00 − $15.00 = $23.00

p. 110   3:00

p. 111

1. 
2. 
3.
4. 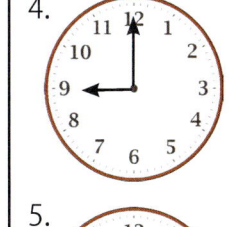
5. 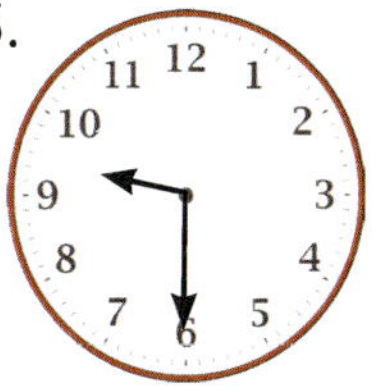
6.
7.

151

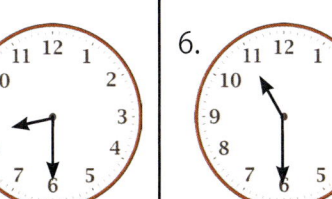

Chapter 7 ● Answer Key

p. 112

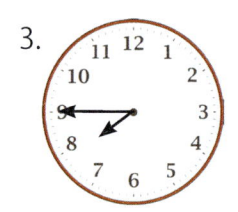

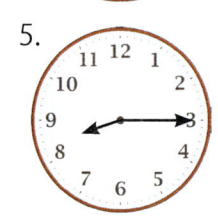

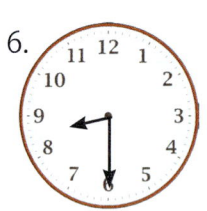

1.
2.
3.
4.
5.
6.

7. 8:45

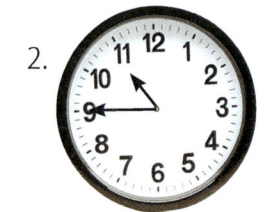

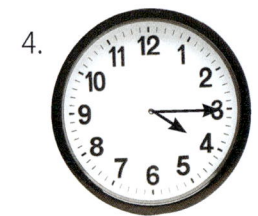

p. 113

1.
2.
3.
4.

p. 114

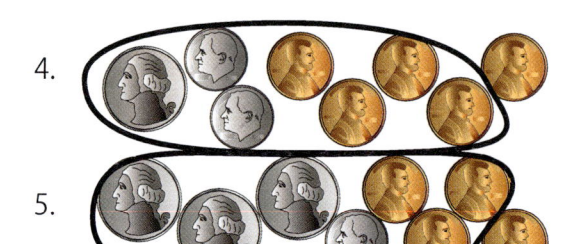

p. 115

Week 1:
Week 2:
Week 3:
Week 4:

p. 116

Bonus: $16.00

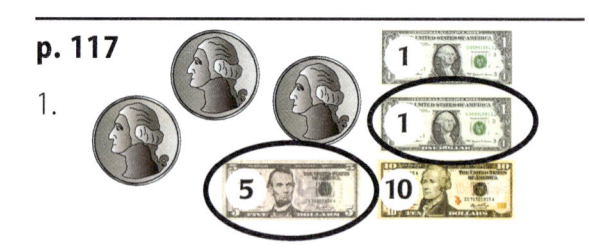

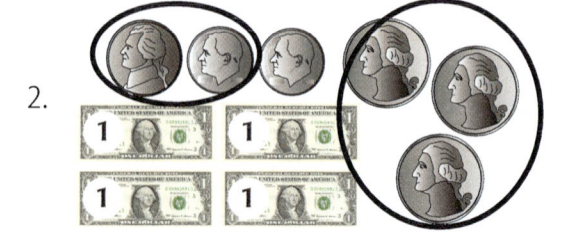

p. 117

3.

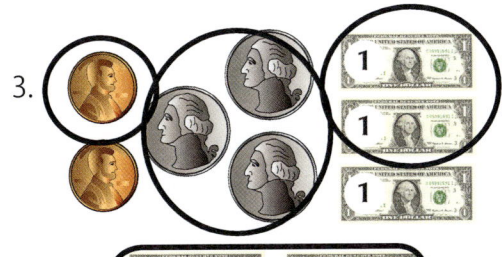

4.

5.

pp. 119

1. $30.00
2. $15.00
3. $25.00
4. $50.00
5. $30.00
6. $20.00

pp. 121

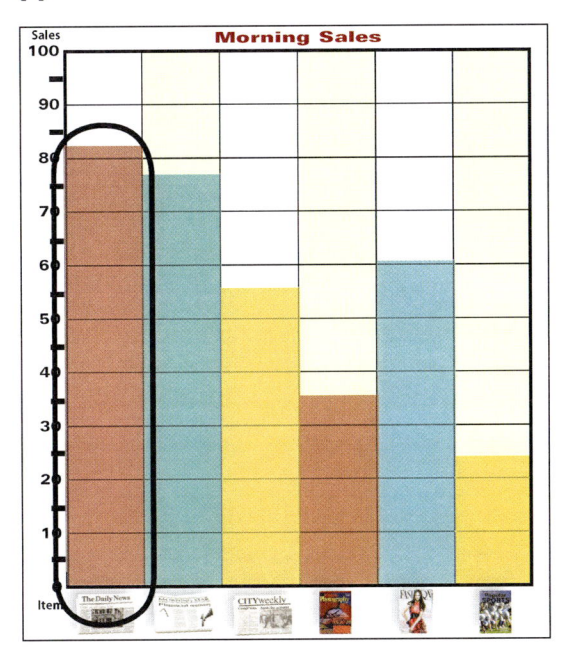

p. 122

Dominick: 1. 176 pts.
2. 24 pts.

Shawna: 1. 180 pts.
2. 20 pts.

p. 123

Dominick   Shawna

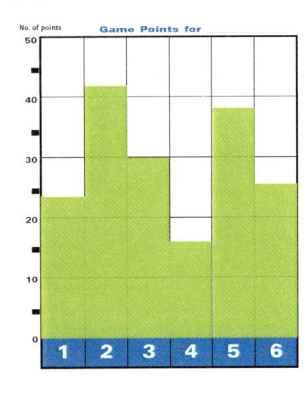

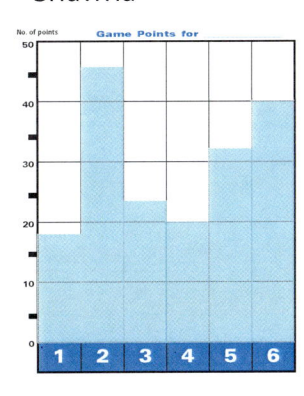

p. 124

1. 22
2. 47
3. 74

p. 125

Chapter 5 • 0–1000

p. 128

Bonus: $25.00

p. 129

Bonus: $82.00

p. 130

Bonus: Lansing

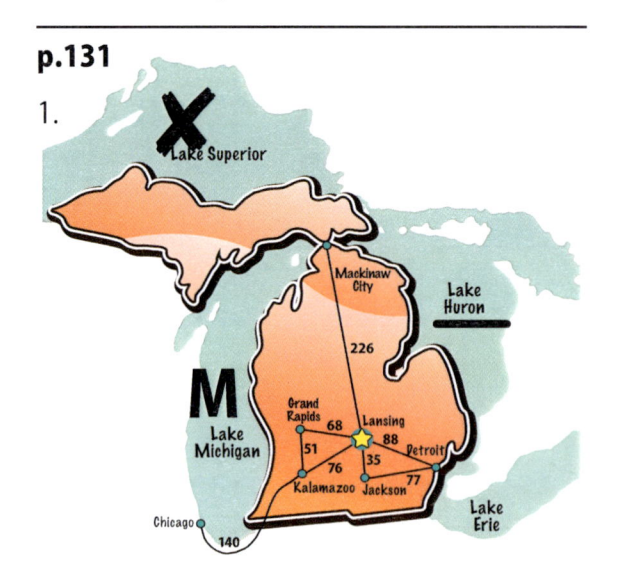

p.131

1.

2. 88 + 68 + **156 mi.**

3. 76 + 140 = **216 mi.**

4. 226 + 35 = **261 mi.**

p. 132

1. 8 + 8 + 8 = **24 ounces**

2. 2 + 2 + 2 + 2 + 2 = **10 pints**

3. 4 + 4 + 4 + 4 = **16 quarts**

Bonus: 2 + 2 + 2 + 2 = **8 pints**

p. 133

125 + 125 = **250 lbs.**

Bonus: 8 + 8 + 8 = **24 pounds**

pp. 135

1. 235 + 312 = **547 lbs.**

2. 320 + 244 = **564 lbs.**

3. 291 + 320 = **611 lbs.**

4. 320 + 312 = **632 lbs.**

5. 235 + 244 = **479 lbs.**

Bonus: 291 + 244 + 312 = **847 lbs.**

p. 136

Bonus: $0.00

p. 137

1. $65.25 + $156.00 = **$221.25**

2. $156.00 + $82.50 = **$238.50**

3. $28.25 + $12.00 = **$40.25**

4. a. answers vary   b. $20.00

Bonus: $70.00

p. 138

1000 − 750 = **250 cans**
or 750 + **250** = 1000 cans

p. 139

Bonus: $295.00 − $90.00 = **$205.00**

pp. 141

1. 2/1/10 to 2/28/10
2. Gross $408. 67
3. Medicare $5.28
4. Social Security $30.65
5. Net Pay $372.74
6. $408. 67 − $372.74 = **$35.93**

Challenge: $1226.01 − $1118.22 = **$107.79**

p. 142

Challenge: $5835.80

p. 143

1. 3/1/10 to 3/6/10
2. $600.00
3. 40 hours
4. Federal Income Tax $60.00
5. Medicare $8.70
6. Social Security $53.62
7. State Income Tax $9.00
8. $714.90 − $583.58 = **$131.32**

p. 144

1. 
2. 
3. 
4. 
5. 
6.

p. 145

1a. 
1b. 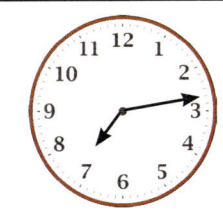
2a. 
2b.

p. 146

1. 7:00
2. 3:30

p. 147

1. $250.00
2. $550.00

p. 149

1. 
2. 
3.

p. 150

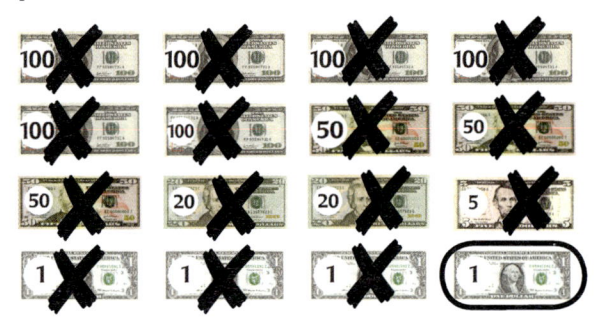

p. 151

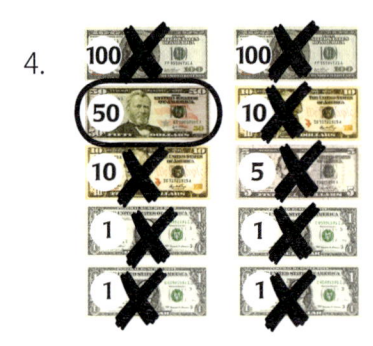

1.
2.
3.
4.

pp. 153

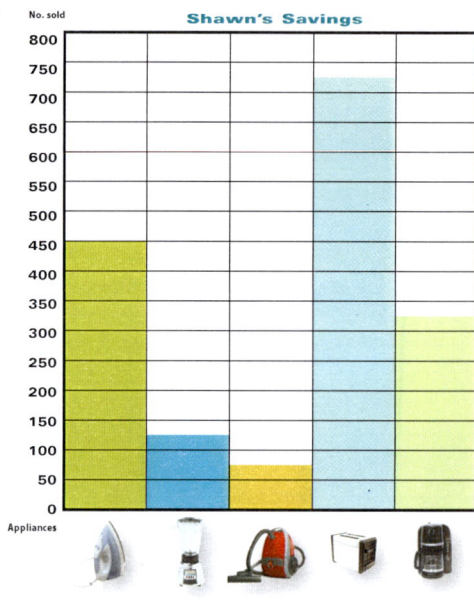

p. 154

1. A – 45 students
   B – 41 students
   C – 40 students
   D – 48 students
2. E – 58 students
   F – 48 students
   G – 52 students
   H – 59 students

p. 155

5. green

R S E D N

Chapter 6 • Fractions

p. 158

= 1/2   Bonus: 1/1

p. 159

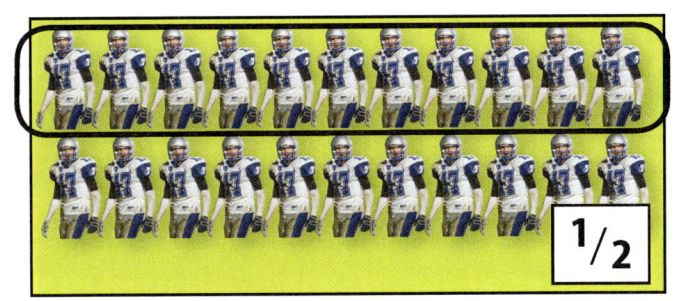

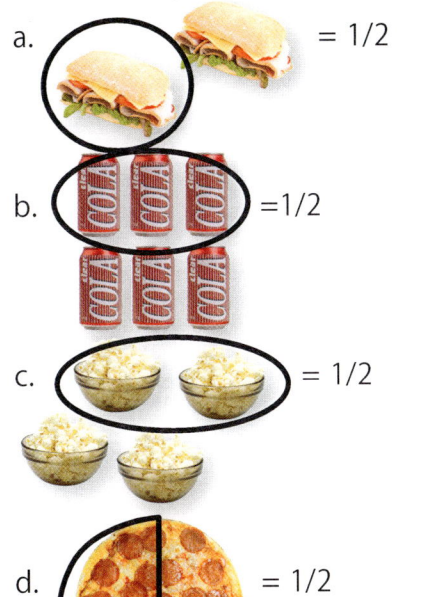

a. = 1/2

b. = 1/2

c. = 1/2

d.  = 1/2

p. 160

0

Bonus:

p. 161

1/4

Bonus:

p. 162

a.

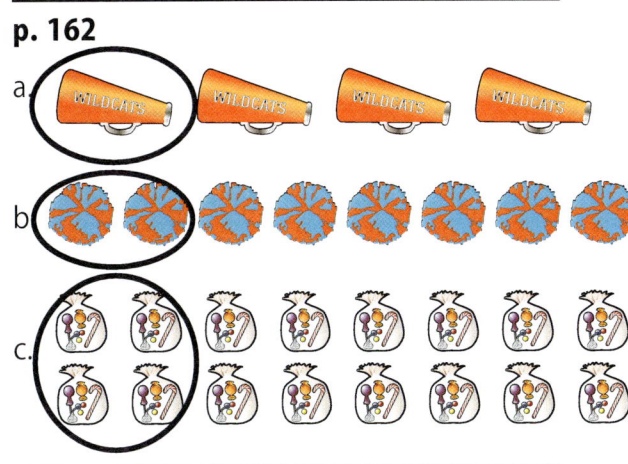

b.

c.

p. 163

1., 2.

3. 1/4

4. 3 cookies

p. 164

= 2/3

p. 165

a. 1/3 of 15 is **5**

b. 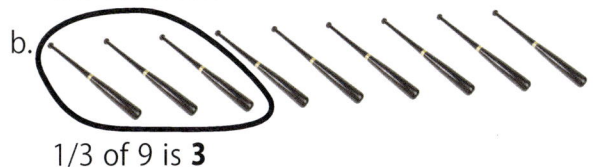 1/3 of 9 is **3**

c.  1/3 of 18 is **6**

d.  1/3 of 3 is **1**

Bonus:

157

Chapter 7 ● Answer Key

p. 166

a. 1/2 c.

b. 1/1

c. 1/3 c.

d. 1/4 c.

e. 2/3 c.

f. 1/4, or 3 eggs

Bonus: (first two quarters circled out of four)

p. 167

1. 2/3
2. 3/4
3. 1/2

p. 168

1/5

Bonus: 2/5

p. 169

1. (4 pennies circled of 5) = 4/5
2. (3 pennies circled of 5) = 3/5

Bonus: (2 pennies circled of 5)

p. 170

1. (2 nickels circled of 5) 2/5
2. (3 nickels circled of 5) 3/5

Bonus: 0 nickels left (all 5 nickels circled)

p. 171

1. (2 of 5 twenties circled) 2/5
2. (3 of 5 twenties circled) 3/5

p. 172

(1 cup measuring cup) = 1/8

Bonus: 6 cards

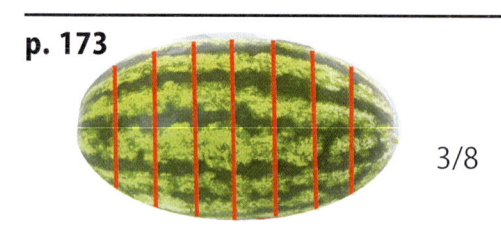

p. 173

(watermelon divided) 3/8

Bonus: 3/4

p. 174

1. 8
2. 1/8
3. 1/8
4. 2/8
5. 3/8
6. 2/8

Bonus: 2/4, or 1/2

p. 175

1/8

Bonus: 4/8, or 1/2

p. 176

(cake) 1/10

Bonus:

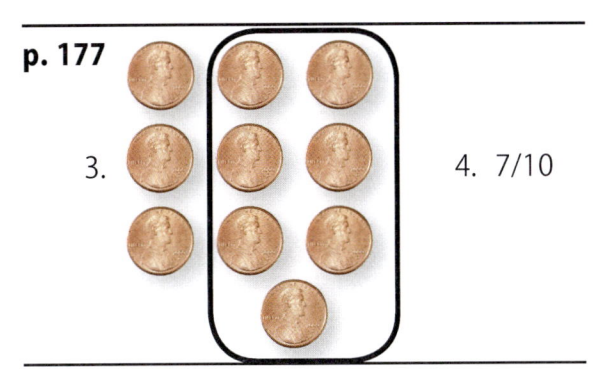

p. 177

3.   4. 7/10

p. 178

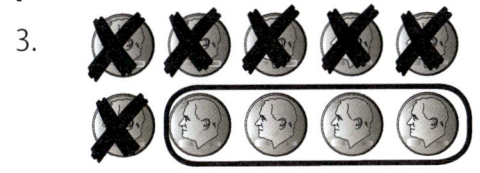

3.

4. 4/10, or 2/5

Bonus:

p. 179

3.

4. 2/10, or 1/5

Bonus:

p. 180

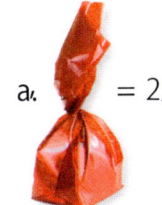

a. = 2/10, or 1/5   b. = 5/10, or 1/2   c. = 2/10

p. 181

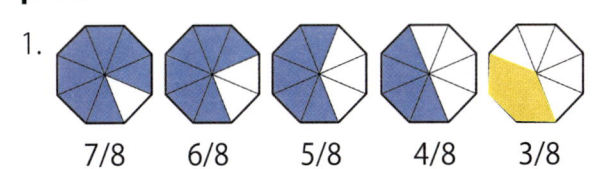

1.

7/8   6/8   5/8   4/8   3/8

2.

1/5   2/5   3/5   4/5

3.

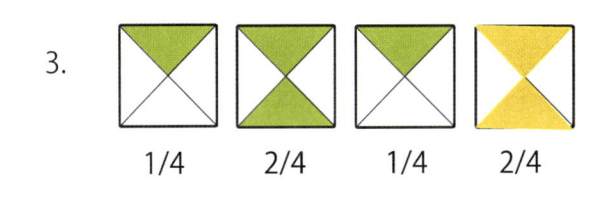

1/4   2/4   1/4   2/4

pp. 183

Answers vary

Chapter 8
Appendix

Vocabulary 1

addition 3  ➕ 4  = 7 	**altogether** 3  ➕ 2  = 5 
behind  ⬇	**between** 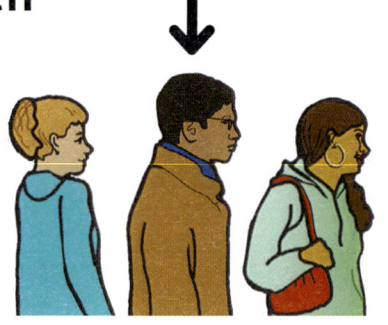 ⬇
counting back 9, 8, 7, 6 …	**counting on** … 5, 6, 7, 8
difference 8 ➖ 4 = 4 	**earlier** (one hour)

Vocabulary 2

equals	first
$6 + 5 \boxed{=} 11$ $7 - 3 \boxed{=} 4$	

(how many) **in all**	**in front of**
$5 \boxed{+} 2 = 7$	

label	**last**
$8_{\text{points}} + 10_{\text{points}} = 18_{\text{points}}$	

later	**least**

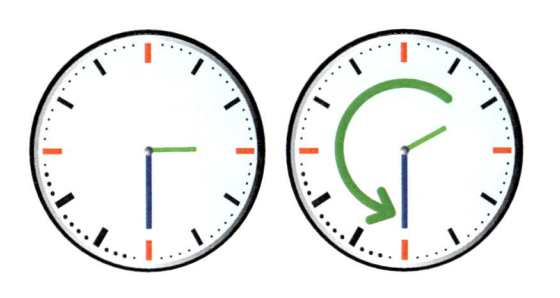

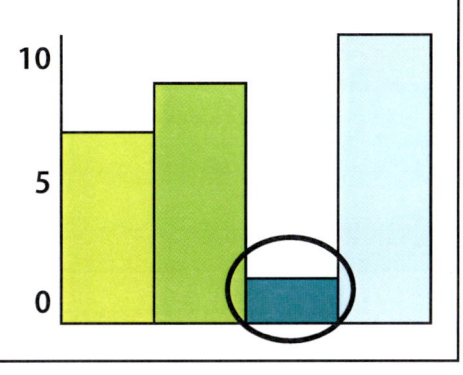

▼ Note: When the words "how many more" appear in the problem, count on or count back.

Chapter 8 ● Appendix

Vocabulary 3

left	minus
	−

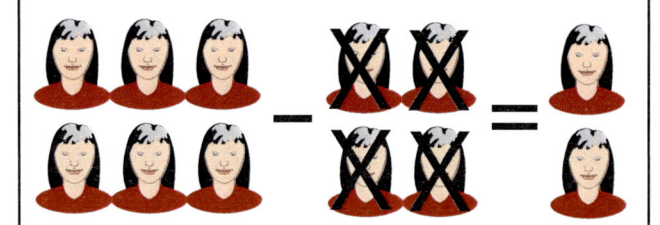

most	plus
	+

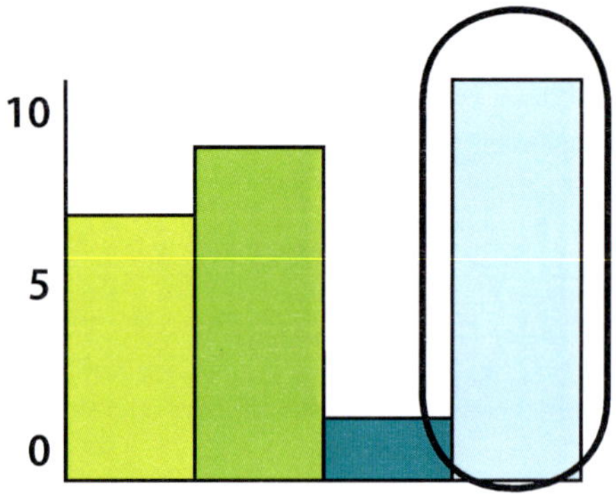

remainder	subtraction
$12 - 9 = 3$	−

sum	total amount
$14 + 3 = 17$	

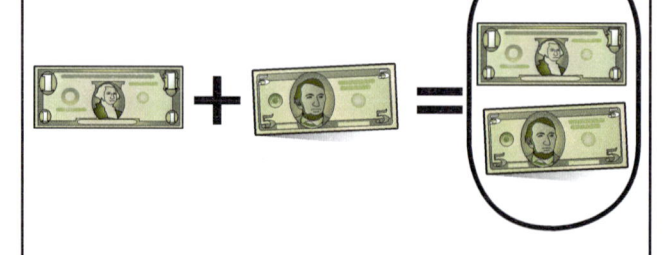

Vocabulary 4

word problem

Kevin has 4 [dollar bill].

He earned 3 more [dollar bill].

How many [dollar bill] did Kevin earn in all?

▼ Note: Use the blank cards for additional words or to make your own cards.

One Hundreds Chart

1	2	3	4	5	6	7	8	9	10
11	12	13	14	15	16	17	18	19	20
21	22	23	24	25	26	27	28	29	30
31	32	33	34	35	36	37	38	39	40
41	42	43	44	45	46	47	48	49	50
51	52	53	54	55	56	57	58	59	60
61	62	63	64	65	66	67	68	69	70
71	72	73	74	75	76	77	78	79	80
81	82	83	84	85	86	87	88	89	90
91	92	93	94	95	96	97	98	99	100

Bibliography

Attainment's Monthly Calendar. Attainment Company, Inc., 2007.

Attainment's Weekly Planner. Attainment Company, Inc., 2007.

Kinney, Judi. *Adapting Math Curriculum: Money Skills.* Attainment Company, Inc., 2003.

Kinney, Judi and Fischer, Debbie. *Differentiated Math Lessons.* Attainment Company, Inc., 2009.

Moore, Jo Ellen. *Daily Word Problems.* Evan-Moor Educational Publishers, 2001.

Pisano, Sal. *Measurement 1.* Rosen Publishing Group, Inc., 2004.

Strazzabosco, John. *Measurement 3.* Rosen Publishing Group, Inc., 2004.

Van Leeuwen, Michele. *"Real Life" Math Word Problems*, 2005.